ENDOCRINOLOGY RESEARCH AND CLINICAL DEVELOPMENTS

DYSLIPIDEMIA: CAUSES, DIAGNOSIS AND TREATMENT

Endocrinology Research and Clinical Developments

Additional books in this series can be found on Nova's website under the Series tab.

Additional E-books in this series can be found on Nova's website under the E-book tab.

ENDOCRINOLOGY RESEARCH AND CLINICAL DEVELOPMENTS

DYSLIPIDEMIA: CAUSES, DIAGNOSIS AND TREATMENT

MIROSLAVA KARAPETROVIČ
AND ZLATKO AČIMOVIČ
EDITORS

Nova Science Publishers, Inc.
New York

For permission to use material from this book please contact us:
Telephone 631-231-7269; Fax 631-231-8175
Web Site: http://www.novapublishers.com

LIBRARY OF CONGRESS CATALOGING-IN-PUBLICATION DATA

Dyslipidemia : causes, diagnosis, and treatment / editors, Miroslava Karapetrovic and Zlatko Acimovic.
p. ; cm.
Includes bibliographical references and index.
ISBN 978-1-62100-917-7 (hardcover)
I. Karapetrovic, Miroslava. II. Acimovic, Zlatko.
[DNLM: 1. Dyslipidemias. 2. Lipoproteins, HDL--physiology. WD 200.5.H8]

572.68--dc23

2011040279

Published by Nova Science Publishers, Inc. † New York

CONTENTS

PREFACE

Dyslipidemia is an abnormal amount of lipids in the blood. In this book, the authors present topical research on the causes, diagnosis and treatment of dyslipidemia. Topics discussed include diet and nutrition therapy in dyslipidemia management; fish proteins for coronary artery disease; qualitative and quantitative characteristics of low-density and endothelial lipase and its role in the metabolism of HDL-cholesterol.

Chapter 1 - Abnormal blood lipids is a condition of elevated cholesterol, low-density lipoproteins, triglycerides or low high density lipoproteins leading to hypercholesterolemia with risk of atherosclerosis or diabetes. Status of dyslipidemia changes in different age groups and ethnicities in different life styles. Dietary low fat intake or nutrition supplementation is the key of lowering blood lipids as art for keeping dyslipidemia in control and manageable. Use of wild and traditional foods is less known in keeping good health at low risk of dyslipidemia or cardiac diseases. Ancient view of low fat, vegan dietary intake, exercise and active life style puts forth evidence of long life expectancy at low risk of cardiovascular disease as observed among tribals in present times. Extensive clinical trials suggest ambiguous state of strictly medical treatment of dyslipidemia to completely cure the disease while it causes more side effects. Recent clinical trials show clear benefits of dietary and nutrition therapy in lipid lowering and management of dyslipidemia in order to keep cholesterol in normal range. New guidelines suggest the search of new more sensitive biomarkers of dyslipidemia or cardiovascular disease to assess the disease growth, need of active life style, better knowledge of lipoprotein metabolic control to fine tune the dosage and selection of drugs combined with dietary or nutraceutical regimens under medical supervision. Ancient traditional ways of active life style suggest benefits of 'spiritual acceptance' or no affluence, no use of processed foods or discipline of regular fasts to keep low fat diet and good health with high life expectancy.

Chapter 2 - It is recognized that substantial cardiovascular (CV) risk remains even after lowering LDL-C, and that low HDL-C is associated with this residual risk. The anti-atherogenic role of HDL-C is supported by its ability to mediate reverse cholesterol transport, reduce oxidation and inflammation. This has led to substantial interest in HDL-C as a therapeutic target. However, several lines of evidence suggest that HDL-C concentration may be an inadequate representation of its anti-atherogenic effect. Considerable variation is found in cholesterol efflux capacity, a measure of HDL function, amongst individuals with the same HDL-C levels. There have also been situations in which increased CV risk was not observed in individuals with low HDL-C, such as in certain genetic conditions of low HDL-C and ApoA-1. These observations highlight the limitations of HDL-C concentration as a surrogate for anti-atherogenecity. It also points to a need for a more refined measure for use as a therapeutic endpoint. Our ability to quantify the anti-atherogenic effects of HDL-C more accurately will become increasingly important as more HDL-C based therapies are developed. There are two proposed approaches to address the limitations of measuring HDL-C levels. The first is an assessment of HDL-C heterogeneity. HDL-C exists as a heterogeneous population of subfractions of different densities, charges, sizes and apolipoprotein compostion. These subfractions have been found to correlate differently with CV risk. The second approach is to quantify HDL-C function, using measures such as cholesterol efflux capacity and anti-inflammatory activity. In this chapter, we will review the existing methods of evaluating HDL-C heterogeneity and function. We will also discuss their limitations and potential clinical applications in terms of refinement of cardiovascular risk prediction, and assessment and monitoring of response to HDL-based therapies.

Chapter 3 - Cardiovascular disease (CVD) is leading cause of morbidity and worldwide. Higher levels of low-density lipoprotein cholesterol (LDL-C) are associated with an increased risk of coronary heart disease (CHD), myocardial infarction, and stroke. Guidelines of the Adult Treatment Panel III emphasize intensive reduction of LDL or non-high-density lipoprotein cholesterol in patients at high risk of CHD. Restriction of dietary saturated and trans-fat and cholesterol, along with increased intake of soluble fiber, can achieve LDL-C lowering. However, pharmacological treatment is needed in most patients. Statins are highly efficacious as LDL-C lowering agents and have more modest effects on very LDL-C, triglycerides and HDL-C levels.

The effects of statins on plasma lipids result from their ability to both increase the efficiency with which very LDL-C and LDL-C are cleared from the circulation and reduce the production of apoB-containing lipoproteins by the liver. Placebo-controlled intervention studies of statin drugs for lowering LDL-C provide clear evidence of CVD prevention. In addition, improvement of survival with statins may be due to other pleiotropic effects beyond LDL-C lowering. Side-effects, myopathy and hepatotoxicity, are infrequent and usually mild, but use of lipid-modifying medication demands caution because of the possibility drug interactions. Many recent studies suggest that low levels of HDL-C are a major independent risk factor for CVD. According to several clinical trials, a 1% increase in HDL-C is associated with a 0.7%–3% decrease in CHD events. The direct link between high levels of triglycerides and CHD is less well defined. Fibrates reduce triglycerides levels and may be beneficial for CHD prevention. Fibric acid derivatives and niacin are primarily used to increase HDL-C levels, although with side effects. Low-dose n-3 fatty acids could be used routinely after a myocardial infarction, but the value of higher doses of n-3 fatty acids in reducing cardiovascular risk remains to be demonstrated. In conclusion, achieving low lipid levels appears safe, but the generalizability of these findings to broader populations and the clinical benefit on the reduction of cardiovascular complications remains to be proven.

Chapter 4 - Recent recommendations of federal and regulatory agencies suggest revisit the claims of fish oils rich in omega-3 fatty acids and benefits in prevention of cardiovascular disease. We propose the importance of fish proteins in prevention of CVD to offer better opportunity of lowering blood pressure and coronary artery disease burden. Fish proteins are analyzed by protein electrophoresis after extraction from fish. The nutrition value of fish protein components are estimated and compared with recommended daily allowances (RDA). However, benefits of fish proteins remain less known due to poor knowledge of clinical outcomes, endpoint calibrations and their food value. The current status of fish dietary research and possibility of their use in coronary artery disease as cardioprotective food is presented if the fish protein intake may be better approach over omega fatty acids as reliable, relevant dietary recommendations for CVD in future.

Chapter 5 - Elevated serum low-density lipoprotein cholesterol (LDL-cholesterol) concentration is firmly established as a risk factor for atherosclerosis and cardiovascular disease (CVD). Yet, a number of CVD patients have LDL-cholesterol levels within the recommended range, suggesting the need for advanced lipid testing to determine residual risk. In this manner, further improvement of risk assessment might be accomplished through a more detailed insight into qualitative and quantitative characteristics of LDL particles. Plasma LDL population comprise complex spectrum of particles of different size, density and lipid composition. There is now ample of evidence that certain LDL subclasses, particularly small, dense LDL particles, are superior to LDL-cholesterol in terms of CVD risk prediction. Furthermore, elevation of small, dense LDL particles is also observed in various atherosclerosis-related conditions, such as end-stage renal disease or ischemic stroke. Whereas the measurement of LDL-cholesterol concentrations has proven clinical utility, the usefulness of LDL particles characterisation in clinical practice needs to be further explored. Understanding of structural complexity of LDL particles and their functional consequences might improve prevention and prognosis of atherosclerosis-related diseases, leading toward more specialized therapeutic approaches. Therefore, the questions of whom, when and how to asses small, dense LDL particles still remain open.

Chapter 6 - It is known that the metabolic syndrome (MS) leads to serious cardiovascular disease which continues to be the number one cause of death in developed countries. MS is characterized by dyslipidemia, hypertension, glucose tolerance, and obesity. The MS distribution is growing catastrophically, but molecular mechanisms responsible for developments of complex impairments in MS still remain basically poorly investigated. The formation of complex MS symptoms suggests systemic impairments in lipid and carbohydrate metabolism; it appears that these impairments should have a common basis at the level of expression of appropriate genes. Expression of genes involved into lipid and carbohydrate metabolism is regulated by various transcription factors, including those referred to the superfamily of nuclear hormone receptors: peroxisome proliferator-activated receptors (PPAR), liver X receptors (LXR), pregnane X receptors (PXR), and constitutive androstane receptors (CAR). PXR and CAR are mainly known as the sensors of xenobiotics. Since these transcription factors are ligand-activated, they represent perspective targets for pharmacological treatments. In the recent time, the search of natural and synthetic PPAR ligands is in intensive process. There is a cross-talk between signal transductionpathways of PPAR, LXR, PXR, and CAR; this suggests their integrated role in regulation of genes of lipid and carbohydrate metabolism. In our work at the model of rat strain with Inherited Stress-Induced Arterial Hypertension which demonstrates all signs of MS development it was shown that an increased content of triglyceride, VLDL and LDL cholesterol, a decreased content of HDL cholesterol, a high level of apoB-100, and a decreased level of apoA-I, increased body weight and blood glucose level are associated with increased DNA-binding activity of PPAR,

LXR, PXR, and CAR in the liver; this suggests involvement of these transcription factors in the development of MS in ISIAH rats. We have also shown that stress can affects the DNA-binding activity of these transcription factors. In this review the role of ligand-activated transcription factors PPAR, LXR, CAR, PXR and the role of stress in the development of metabolic syndrome will be discussed.

Chapter 7 - The endothelial lipase (EL) is an enzyme with lipolytic activity against phospholipids of high density lipoprotein-cholesterol (HDL) in plasma. The EL is synthesized by vascular endothelial cells. In particular, it is synthesized mainly in heart, muscle, and adipose tissue and then transported to the luminal surface of endothelial cells where it hydrolyzes lipoprotein triglycerides and phospholipids. In addition, the endothelial-derived EL can be a direct regulator of HDL metabolism increasing the catabolism of HDL particles, thereby reducing the levels and the size of the tank and impacting recycling. B Both experimental in vitro and in vivo studies have revealed remarkable information with regard the physiological role of EL in inflammatory conditions and its potential role in modulating lipoprotein metabolism under inflammatory conditions, including atherosclerosis. EL also seems to have local actions in the vascular wall and the regulation of circulating lipoproteins, and it seems to be associated with established and novel cardiovascular risk factors like hypertension, diabetes, dyslipidemia, obesity, and leptin, which potentially contributes to the progress of atherosclerosis and therefore could be regarded as a biochemical marker of atherosclerosis. The EL inhibition is considered to be an attractive intervention target in order to favorably modulate the atherosclerosis process by improving the HDL levels. This article reviews the effect of EL on the metabolism of HDL-C, its association with cardiovascular risk factors and the potential perspective to arrest or even improves atherosclerosis by EL inhibition.

Chapter 8 - Microvascular damage is an important pathophysiologic mechanism underlying peripheral and autonomic neuropathy. Microvascular injury is mediated by various vascular risk factors, which are also components of the metabolic syndrome. There is increasing evidence for the role of dyslipidemia in the development and progression of diabetic peripheral and autonomic neuropathy. Although small fiber neuropathy has been reported in non-diabetic patients with hypertriglyceridemia, the role of dyslipidemia in peripheral neuropathy is often examined in the context of other vascular risk factors which may confound the association. A number of long-term studies of large cohorts of patients with diabetes have been completed. In these studies over the course of seven or more years of follow-up, it is not glycemia but serum lipids, hypertension, and body mass index that are independently associated with the risk of developing diabetic neuropathy. There is emerging evidence that the aggregate effect of all the vascular risk factors may induce microvascular damage resulting in peripheral neuropathy. Oxidative stress provoked by oxidized LDL amongst other mechanisms may play a major role and constitute a therapeutic target. Statins and fibrates may play a role in the prevention and management of diabetic peripheral neuropathy.

Chapter 9 - *Objectives:*Fenofibrate exerts multiple pleiotropic effects as a selective peroxisome proliferator-activated receptor alpha (PPAR-α) agonist in addition to its well-characterized lipid-lowering effects. Fenofibrate suppressed the production of pro-inflammatory cytokines through the inhibition of NFκB pathway and ubiquitin-proteasome system, leading to the decrease in the incidence of atherosclerosis and type II diabetes mellitus (DM) that are tightly associated with chronic inflammation. Anti-inflammatory and anti-oxidant activities of fenofibrate have been studied in the chronic inflammatory diseases such as rheumatoid arthritis (RA) with a higher risk of atherosclerosis-related death. *Methods:* In this chapter, our long-term follow-up study on the anti- inflammatory and anti-oxidative activities of fenofibrate in RA were compared with the parallel study of statins therapy and control concerning to the changes in the serum levels of lipid profile and inflammatory parameters including C-reactive protein (CRP), erythrocyte sedimentation rate (ESR), matrix metalloproteinase-3 (MMP-3), pentosidine and homocysteine. *Results:* Both serum lipid profile and the most inflammatory parameters such as CRP, ESR and MMP-3 remained basically stable in all the 3 groups during the 18 months study. Although serum pentosidine and homocysteine levels elevated significantly in the control group without lipid-lowering agents, the two parametersdid not change significantly in fenofibrate and statins groups. Prednisolone dosage decreased significantly in all the 3 groups. *Conclusions:*Lipid-lowering agents such as fenofibrate and statins may play a role as anti-atherogenic and anti-oxidant agents in the chronic inflammatory disease like RA.

In: Dyslipidemia
Editors: Z. Ačimovič et.al, pp. 1-52
ISBN: 978-1-62100-917-7

Chapter 1

DIET AND NUTRITION THERAPY IN DYSLIPIDEMIA MANAGEMENT

Rakesh Sharma[1] and Robert J. Moffatt[2]
[1]Amity Institute of Nanotechnology,
Amity University Uttar Pradesh, NOIDA, India
[2]Department of Exercise, Food and Nutrition,
Florida State university, Tallahassee, FL 32304, USA

ABSTRACT

Abnormal blood lipids is a condition of elevated cholesterol, low-density lipoproteins, triglycerides or low high density lipoproteins leading to hypercholesterolemia with risk of atherosclerosis or diabetes.

Status of dyslipidemia changes in different age groups and ethnicities in different life styles.

Dietary low fat intake or nutrition supplementation is the key of lowering blood lipids as art for keeping dyslipidemia in control and manageable.

Use of wild and traditional foods is less known in keeping good health at low risk of dyslipidemia or cardiac diseases.

Ancient view of low fat, vegan dietary intake, exercise and active life style puts forth evidence of long life expectancy at low risk of cardiovascular disease as observed among tribals in present times.

Extensive clinical trials suggest ambiguous state of strictly medical treatment of dyslipidemia to completely cure the disease while it causes more side effects.

Recent clinical trials show clear benefits of dietary and nutrition therapy in lipid lowering and management of dyslipidemia in order to keep cholesterol in normal range.

New guidelines suggest the search of new more sensitive biomarkers of dyslipidemia or cardiovascular disease to assess the disease growth, need of active life style, better knowledge of lipoprotein metabolic control to fine tune the dosage and selection of drugs combined with dietary or nutraceutical regimens under medical supervision.

Ancient traditional ways of active life style suggest benefits of 'spiritual acceptance' or no affluence, no use of processed foods or discipline of regular fasts to keep low fat diet and good health with high life expectancy.

Keywords: Dietary management, dyslipidemia, cholesterol, nutrition, lipid

1. Introduction

Dyslipidemia is an abnormal amount of lipids (e.g. cholesterol lipid, lipoprotein, or apolipoprotein levels) in the blood. The word 'dyslipidemia' was coined way back in 1965 by Fredrickson as major seven classes of elevated lipids in the body [1]. Lipid disease initiates if dyslipidemia goes uncontrolled at biochemical level with result of increased glucose (diabetes) or lipid and lipoprotein deposits (obesity and hypercholesterolemia). It is believed that continuous lipid profile monitoring at regular intervals and timely effective intervention (exercise, dietary, nutrition supplements, lipid lowering foods and behavioral counseling) keeps dyslipidemia in control and it is manageable for longer life expectancy when pharmacological medication fails or shows side effects [Dowman]. In last decade, management of dyslipidemia has remained a major controversy to meet target levels of LDL-C, triglycerides and keeping up HDL-C in blood for long time by antihypertensive therapy, statins, niacin, fibrates, omega-3 fatty acids, and dietary management especially in children, adolescents and elderly population [2]. With all these factors in consideration, present chapter focuses on dyslipidemia in following sections including definition, lipids lipoproteins, concepts of lipid disorder and lipid lowering, ancient and modern views of low fat diets, nutrition therapy in clinical trials, less known low fat risk tribal life styles, guidelines and futuristic prospects of dyslipidemia management. Our ultimate message is "Effective dyslipidemia management by 'non-drug therapy and effective intervention' prevents person from cardiovascular disease". The present chapter puts forth evidence of traditional non-affluent and ancient life styles as favorable to keep good health.

Definition of Dyslipidemia

Dietary fat-intake and elevated blood lipids are considered as risk of dyslipidemia and development of atherosclerosis and diabetes. The elevated cholesterol and lipoproteins are the main lipids to cause dyslipidemia. The elevated lipids are known as triglycerides (TC), lipoproteins (LDL-C), apolipoproteins (apo-B), non-HDL-C, and TG and low HDL-C and apo A-I in the blood. The lipid disorders are classified into seven dyslipidemia profiles [1].

- Elevated LDL-C (type IIa)
- elevated LDL-C combined with high triglyceride (TG) (type IIb)
- elevated TG (type IV)
- low HDL-C (hypo-α),
- elevated LDL-C type IIa, type IIb, or type IV accompanied by low HDL-C;
- hyper-apobetalipoproteinemia (hyper-apoB), *i.e.* elevated apolipoprotein B (apoB) but normal LDL-C
- Inherited lipoprotein disorders that often present in youth at high risk of future CVD include familial hypercholesterolemia (FH), familial combined hyperlipidemia (FCHL), hyper-apoB, familial hypoalphalipoproteinemia, apolipoprotein A-I mutations, common and rare variants in ABCA1 including Tangier disease, lecithin cholesterol acyl transferase (LCAT) deficiency, and hyper-TG associated with lipoprotein lipase deficiency and defective apoC-II. The disorder of cholesteryl ester transfer protein deficiency, often presents as high HDL-C, but its increased or reduced risk of CVD is not resolved [2].

Lipids and Lipoproteins in Dyslipidemia

In blood, lipids are transported in almost all organs and remain in circulation to compensate the energy demands. By density gradient centrifugation of plasma, different lipid fractions can be separated out in chloroform-methanol organic phase as shown in Figure 1 [1].

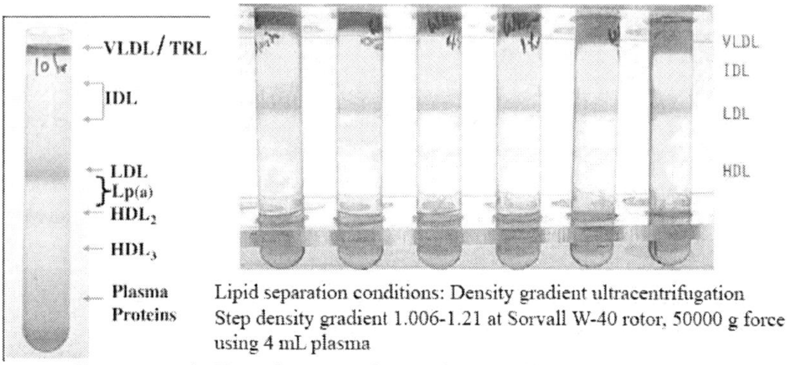

Source: Sharma et al. How doctor evaluates the carotid artery disease in patients? http://www.scribd.com/doc/24292286/Evaluation-of-Carotid-Artery-Disease.

Figure 1. A set of lipids is shown after centrifugal separation into VLDL, IDL, HDL. The measurement of different lipids can measure Cholesterol: HDL and IDL fractions to confirm the dyslipidemia. The elevated lipids predict a first guideline of disease progress evaluation and response of drug therapy.

Different lipid fractions and lipoproteins can be measured to evaluate the dyslipidemia and possibility of cardiovascular risk. Both hypercholesterolemia and metabolic syndrome are associated with atherogenic and diabetes related dyslipidemia, which is characterized by high TG, elevated small dense LDL-C and low HDL-C. Such dyslipidemia poses a high risk of CHD on patients. According to NCEP ATPIII, elevated TG is measured as very LDL (VLDL) as marker for atherogenic remnant lipoproteins or non-HDL-C6. Furthermore, NCEP III guidelines state that non-HDL-C is calculated by subtracting HDL-C from total cholesterol and should be a secondary target if TG is >200 mg/dL. Atherogenic dyslipidemia (LDL-C, HDL-C, and TG) are interrelated and each component predicts CHD risk.

Low-Density Lipoprotein Cholesterol

LDL-C is a combination of lipoprotein (LDL) and lipid-like cholesterol. Cholesterol is circulated in blood as free or cholesterol esters along with bound form with lipoproteins in the form of LDL-C. LDL-C can be small dense LDL particles can be more in number than blood cholesterol. However, LDL-C value measurement in a standard lipid profile does not provide any information about the size of LDL particles [5].

High-Density Lipoprotein Cholesterol

HDL is a high-density lipoprotein to transport cholesterol back to liver. HDL-C consists HDL particle with cholesterol ester inside. Low HDL-C is risk factor for CHD (golden rule is that every 1 mg/dL increase in HDL decreases 2% to 3% risk of CHD). Although the mechanism is not clear, it is believed that the anti-atherogenic effect of HDL-C may be a result of reverse cholesterol transport (RCT), and anti-oxidant and anti-inflammatory properties as shown in Table 1. However, small size of HDL-C particles and action of CETP (cholesterol ester transfer protein from liver) play an important role to make small HDL-C particles by exchanging more TG from VLDL particle and cholesterol esters from HDL-C. Smaller HDL-C particles are readily excreted out from the kidneys resulting in lower LDL-C particles [6].

Table 1. The multiple anti-atherogenic protective actions of high-density lipoprotein cholesterol

1. Reverse cholesterol transport
2. Cellular cholesterol efflux
3. Anti-inflammatory action
4. Anti-infectious
5. Anti-oxidative
6. Anti-thrombotic
7. Anti-apoptotic
8. E ndothelial repair
9. Vasodilatory activity

In following section, we describe individual lipoproteins.

Triglycerides

Triglycerides are small lipid molecules. They are circulated in blood in exchange with VLDL particles and constitute the lipoproteins. The relationship of LDL and TG to composite endpoints of CHD was assessed in Pravastatin or Atorvastatin Evaluation and Infectious Therapy-Thrombolysis in Myocardial infarction (PROVE IT-TIMI)-22 trial [7]. TG < 150 mg/dL was associated with lower CHD risk in ACS patients.

Apart from lipoproteins, other factors such as homocyst(e)ine, C-reactive protein, genetic risks and hormone replacement therapy also influence cardiovascular disease risk.

What Is Lipid Lowering?

The elevated blood lipids are the result of lipid disorder(s) in the body mainly due to interralated lipid metabolic changes in liver, intestine, adipose tissue at the cellular level of membrane, mitochondria and macrophages as shown in Figure 2.

The central regulatory pathway is cholesterol biosynthesis and its breakdown into bile acids. The first regulatory enzyme segment includes 3-Hydroxy-3-Methylglutaryl Coenzyme A (HMG-CoA) reductase (HMG-CoAR for cholesterol synthesis) and Cholesterol-7α Hydroxylase (CHOL-H for cholesterol breakdown) to precursors. The ratio of HMG-CoAR:CHOL-H plays a significant role in control of lipids in the body. The 'lipid lowering' or cutting down cholesterol is the measure of bring back the elevated blood lipids to normal lipid levels in the body by using the choice of HMG-CoAR inhibitors (mainly statins) or CHOL-H stimulators (mainly bile acid sequesterants). The lipid lowering may also include dietary low fat intake or synthetic diets to keep low supply of lipids by utilizing the stored lipids and fats (Step I and Step II diets). However, broadly success in lipid lowering by use of diets or drugs also depends on several other body constitutional, age, heredity, socio-economic, and environmental factors.

How Severe Is Dyslipidemia to Cause Lipid Disorders?

The elevated lipids, lipoprotein, and apo-B levels are common in primary dyslipidemias associated with premature CAD and CVD. FH is an autosomal dominant disorder due to defects in the LDL receptor (LDLR) gene (see Figure 1). FCHL and hyper-apo-B result from overproduction of VLDL, IDL, and LDL (see Figure 1).

The expression of FCHL can be delayed or associated with type IIa, IIb, or IV lipoprotein profiles or isolated high apoB with premature CAD. The apoB and apoA-I apolipoprotein levels were stronger predictors of parental CVD than LDL-C and HDL-C.

National Health and Nutrition Education Survey (NHANES) in collaboration with NCEP and AHA have set guidelines on use of lipid profiles in evaluation of dyslipidemia.

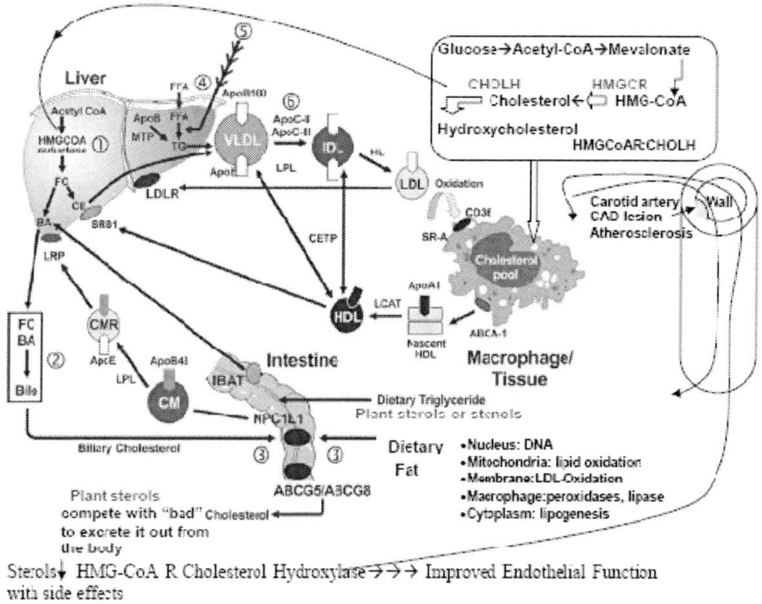

Figure 2. Three major physiological pathways of plasma lipoprotein metabolism are shown: 1) transport of dietary (exogenous) fat (*left*), 2) transport of hepatic (endogenous) fat (*center*), and 3) reverse cholesterol transport (*bottom*). Sites of action of the six major lipid-altering drugs on exogenous and endogenous pathways of lipoprotein metabolism are 1) inhibition of HMG-CoA reductase by statins and dependence on HMG-CoAR:Choleterol hydroxylase; 2) binding of bile acids by sequestrants, interfering with their reabsorption by the ileal bile acid transporter (IBAT); 3) binding of a cholesterol absorption inhibitor to the Niemann Pick C1L1, decreasing the absorption of dietary and biliary cholesterol; 4) Decreased mobilization of free fatty acids (FFA) by niacin, leading to decreased uptake of FFA by liver and reduced VLDL, IDL, and LDL production; 5) inhibition of TG synthesis by ω-3 fatty acids; 6) up-regulation of lipoprotein lipase (LPL) and decreased production of apoC-III, an inhibitor of LPL, by a fibric acid derivative, leading to decreased VLDL-TG. The hepatic cholesterol pool is decreased by the agents at steps 1, 2, and 3, each leading to an up-regulation of the LDLR. In tissues, the lipid synthesis and oxidation at different cellular organelle plays role shown with dots. LCAT, Lecithin cholesterol acyl transferase; ABCA-I, ATP-binding cassette protein A-I; ABCG, ATP-binding cassette protein G; BA, bile acids; CE, cholesteryl esters; CM, chylomicrons; CMR, chylomicron remnants; SR-A, class A scavenger receptor; SRB1, class B scavenger receptor. (Reproduced with permission from Reference [4]).

The lipid profile includes the following lipids including TC, TG, LDL-C, HDL-C, and non-HDL-C. Each lipid is measured by following calculations:

- LDL-C = TC – (HDL-C + TG/5).
- TG = (VLDL)-C. For TG = 400 mg/dl, a direct LDL-C is measured.
- TC, HDL-C, and non-HDL-C are measured by clinical chemistry on nonfasting condition.
- apoB and apoA-I are mesured by immunochemical methods
- Non-HDL-C = TC - HDL-C. It includes apoB-containing lipoproteins [VLDL, intermediate-density lipoprotein (IDL), LDL, and lipoprotein(a)].
- VLDL, LDL, and HDL subclasses by nuclear magnetic resonance spectroscopy or vertical-spin density-gradient ultracentrifugation.

Another set guideline on dyslipidemia management and cardioprotection highlights global risk and several factors influencing risk based on LDL-C (mmol/L) and total colesterol:HDL-C ratio target levels in different risk catagories (<2.5 and <4 at high risk; <3.5 and <5 at moderate risk; and <4.5 and <6 at low risk). [3]. In our study, a cross sectional data showed variation in serum lipid and lipoproteins as risk factors in both sexes in their adolescence, mid age and advanced age.

Table 2. Serum lipid levels and lipoproteins are shown in different age groups of adolescents, mid-age, and advanced age population in a cross section study* Reproduced with permission from reference [113]

	Adolescent			Mid-age			Advanced age		
	Male		Female	Male		Female	Male		Female
Total cholesterol	↓	<<<?	↓	↑	<<< ?	↑	↓↔		↓↔
Triglycerides	↕				>>>		↑	<<<	↑
HDL-C	↓		↕	↓↔	<<<	↑		↕↔	
LDL-C	↕	<<<?	↕	↕	<<<	↑	↓		↓
Lp(a)	↑	>>>	↑	↑	=	↑		NC	
TC + LDL-C	↕	<<<		↑	>>>	↑		NC	
TG + HDL-C	↕	=	↕	↑		↑		NC	

*The data shows relative comparison of serum lipid and lipoproteins with normal values (cut-off 90 percentile): less ↓ ; more ↑; variable ↕; same ↔ ; not done NC; unknown?

2. Historical Sketch of Dyslipidemia: Journey from Ancient to Modern View

2.1. Ancient Concepts of Dietary Habits and Dyslipidemia in Tribes

According to ancient Vedas from India, the span of human civilization has been divided in to four periods, *satyuga, treata, dwaper and kalyuga.* Humans were free from diseases in *satayuga* and the life expectancy was approximately 600 years. In *treata,* the life expectancy reduced to 300 years, in *dwapar* to 200 years and finally in *kalyuga,* it is reduced to 100 years. It is possible that the life expectancy was dependent on physical, social, mental and spiritual health of the human beings. Decrease in physical activity, increase in mental strain, pollution, dietary changes, tobacco consumption, alcoholism and lack of humanity appear to be important contributing factors for decrease in life expectancy in present time [8-14]. Even today, world is unknown to the natural gift of highest life expectancy in India among Indian Hunter-gatherers or Kurichia tribes (16, 17]. We describe here an example of Indian society with less known fact of high life expectancy among animal hunter tribal population and social pattern.

The existing Indian society can be divided in to; primitive hunter-gatherers, peasant-agriculturists and pastoralists, urban slum dwellers and the affluent. The majority of the people arc farmers who are migrating from rural areas to metro city slums or becoming affluent. India and China are countries rich in farmers. Towns have increasing numbers of poor urban slum dwellers and a sizeable affluent society.

From history point of view, dyslipidemia can be explored in diet and lifestyle of Homosapiens and his predecessors Homo erectus and Homo habilis species or 'ancient humans' who were primarily vegetarians. Hunting developed gradually when man moved away from the other primates in competition and got skilled in tool making, while developing social groups; hunting bands and lived in large families (8-14). The man started agro-farming, about 10,000 years ago, with the technical development of agriculture. With increasing food demand, 99% of the time during which man has been evolving from his primate precursors, he remained as hunter-gatherer (8-16). It seems that hunter-gatherers developed a trend of low fat dietary habits. For details description on ancient concepts of nutrition proposed in the Indian scriptures and importance of low fat diet and healthy lifestyle can be read as introduced first time to western literature [18]. In following section, we describe some principles mentioned before [18].

2.2. Ancient Diet Concepts of Dyslipidemia Prevention

The pre-historic concept of diet and nutrition was based on archeological findings in India. Vegetable part of early man's diet or remains of vegetable food was difficult to identify compared to animal bones (10,11). In ancient scriptures of India, dietary patterns such as Sattvic, Rajasik and Tamasic have been described of which the sattvic dietary pattern appears to be similar to modern proposed prudent dietary patterns (Table 3) [12].

Table 3. Low fat and vegan dietary advice in ancient Ayurveda. (extracts from Maha-anaarany Upanishad, 5000 BCE)

1. Sattvic diet. Advised to saints. Contains fruits, vegetables, sprouted grains, roots, tubers, nuts, cow milk, curd, and honey. It would enhance, longevity, health, happiness and spirituality and possible survival may be 100-150 years. Good for mental, social and spiritual health.
2. Rajasic diet. Advised to kings and fighters. Contains fruits, vegetables, nuts, meats from hunted animals, clarified butter, butter, curd, honey, spices, wines. It may cause excitement, confidence and increase intelligence with life expectancy of about 100 years. Good for physical, mental and social health.
3. Tamasic diet. Not advised. Contains high fat fried foods, rich in salt, sugar, spices, chillies, meats from big tamed animals, butter and liquor. This diet makes one dull, enhances anger and criminal tendency and impedes spiritual progress.Life expectancy was much lower. Bad for health.

Other Indian scripture Bhagwatgita (3000 BCE) in following verses suggests the value of vegan and non-meat diet. The first two lines in each of the four verses are given in Sanskrit followed by two lines its translate in English.

Verse: Aayuh satvabalarogyam, sukhpreetiviverchanah, Rasyah snigdhah sthirah hradyah aharah satvikpriyah (Adhyay 27, verse 8).

Translation: The sattvic foods are full of juice, good in taste and increase longevity, wisdom, power, health, happiness, peace and love.

Verse: Yktaharviharasya, yuktachetasya karmasu Yktaswapnavbodhasya, yogo bhavati duhkhaha (Adhyay 6, verse 17).

Translation: Yoga is protective against grief to those who have disciplined diet and lifestyle and controlled bahaviour; sleep and awakening.

Verse: Katvamlalavanatyusnteekshanaruksha vidahinah Ahar rjassyestah, dukhahshokmaypradah, (Adhyay 17, verse 9).

Translation: The foods that are bitter, acid, fried, hot, punjent and dry Give rise to grief, mental stress and diseases.

Verse: Yatyaman gatarasam, pootiparyushitachya yat, Uchhistamapi chamedhyam, bhojanam tamspriyam,(Adhyay 17, verse 10).

Translation: Foods that are cold, prickled, putred and preserved are lured to Criminals and give rise to criminal tendency and behaviour.

Adverse effects of 'dushit'diet and 'haram' lifestyle were also known to great Indian physicians Charaka and Sushruta (600 BCE)[1]. Charaka (600BCE) proposed that, "Heart attack and is born by the intake of fatty meals, overeating, excess of sleep, lack of exercise and anxiety" Charaka Sutra, 600BC. Sushruta being a surgeon, gave a more clear description of atherosclerosis or *madroga* ; "Excess intake of fatty foods and lack of exercise causes obesity and narrowing of the channels taking blood to the heart. It is useful to use guggul, triphala and silajit in the treatment". These herbs are known to have a high content of antioxidant flavonoids, vitamins and minerals as well as fibers. Hippocrates was frequently giving dietary advice to the patients during treatment. Since the days of ancient Greece, doctors in all countries were providing dietary advice to their patients.

In the 7th century BC, a Chinese physician proposed that "increased consumption of salt may cause hardening of the pulse". About 2500 years ago from now(512 BCE), Confucius, the Chinese philosopher taught his students, "the higher the quality of foods, the better and never rely upon the delicacy of cooking". Thus a dietary guideline based on experience, observation and thinking was given as;"cereals, the basic, fruits the subsidiary, meat the beneficial and vegetable as supplementary". On similar lines, WHO experts in

[1] Charaka supposedly lived in Taxila University in the north of India and was a Brahman physician. However, Sushruta was a surgeon from the Vishwamitra family of Varanasi in east of India. In mahabharata, he is represented as a son of that royal sage. The Garuda Puranam places Divodasa as 4th in descent from Dhanavantari, the first propounder of the medical sciences upon earth, whereas the Sushruta samhita describes both as same persons.

1990 proposed the concept of eating a diet high in animal foods, and preference of meat and greasy foods in China. In West, possibly the meat was rich in w-3 fatty acids, and the total fat intake in diet remained within desirable safe limits. Since the first development of agriculture, about 10,000 years ago, stored food became easy practice and major changes in the human diet have occurred in the last 150 years.

2.3. Dietary Fat Intake and Lipid Lowering

In past two decades the major focus was to identify the nutrition factors to promote dyslipidemia and cardiovascular disease mainly higher dietary fat intake exceeding >13% fat energy among almost all socio-economic groups[2]. However, saturated fat and cholesterol rich dietary intake was considered as major contributory factor among effluent societies in western and European world with increased prevalence of heart attack, myocardial infarction, hypertension and atherosclerosis. Continued efforts showed the lipid lowering very effective by dietary intake of polyunsaturated fatty acids, omega-3 and omega-6 fatty acids in cardiovascular prevention in general to keep normal serum lipid profile in both carotid artery and coronary artery diseases[2]. The current trend is to identify endothelial dysfunction of cardiovascular tissue and decrease the inability of heart, coronary and carotid arteries (cardiovascular incapability) to fight against lipid deposition leading to vascular occlusion, stroke and plaque formation by using statins. In this direction, two major suspects of endothelial dysfunction are: 1. dietary fat intake and blood lipids; 2. dysfunction of endothelial cascade and statin side effects. We have focused in present chapter the association of dietary fat intake and blood lipid lowering with detail description on dietary therapy in dyslipidemia.

Dietary Therapy and Lipid Lowering

Initially dyslipidemia are treated with a diet reduced in total fat, saturated fat, and colesterol with aim to lower down LDL-cholesterol and avoid both atherogenesis and metabolic syndrome or diabetes. Two approaches have been proven in this direction. First is intake of saturated fatty acids, trans fatty acids, dietary colesterol, monounsaturated fatty acids, plant sterols/stenols, soluble fibers, soy proteins and micronutrient are suggested as effective supplements

[2] Sharma R. The dietary lipids: The Effect of antihypertensive therapy on serum lipid profile. MSc Applied Nutrition dissertation submitted to National Institute of Nutrition(ICMR), India. 1988;pp 17-33.

to control dyslipidemia. Second approach is intake of nutraceuticals combine with pharmaceutical medical treatment. We describe them in following sections. The NCEP recommended diet treatment after 2 years of age [19]. The STRIP trial recommended low-fat diet at 6 months of age under medical supervisión [20].

Table 4. Acceptable, borderline, and high plasma lipid, lipoprotein, and apolipoprotein concentrations for children and adolescents

Category	Acceptable	Borderline	High[1]	Low[1]
TC	<170	170–199	≥200	
LDL-C	<110	110–129	≥130	
Non-HDL–C	<123	123–143	≥144	
ApoB	<90	90–109	≥110	
TG				
0–9 yr	<75	75–99	≥100	
10–19 yr	<90	90–129	≥130	
HDL–C	>45	35–45		<35
ApoA–I	>120	110–120		<110

Source: National Cholesterol Education Program (NCEP) Expert Panel on Cholesterol. Values for plasma apoB and apoA-I are from the National Health and Nutrition Examination Survey III (NHANES III) (23).

[1] The cutoff points for a high or low value represent approximately the 95th and 5th percentiles, respectively.

Safety and efficacy of dietary therapy in infants, children, and adolescents.

Who Needs Dyslipidemia Treatment with Diet?

- If TC, LDL-C, non-HDL-C, or TG is elevated or HDL-C is low followed by repeat lipid profile 3 wk later to confirm the first lipid profile.

- If one or more of the lipid or lipoprotein values remains above the elevated cutoff point or HDL-C is low, secondary causes of dyslipidemia – need of dietary treatment Step One diet in all ages.
- If repeat lipoprotein profile in 6–8 wk shows dyslipidemia- Step Two diet is initiated in all ages [19].

Human milk remains the gold standard for infant feeding. Nutrition elements such as calcium, zinc, vitamin E, phosphorus supplemented low-fat diet is safe in age of 7 months -11 yr advocated in Special Turku Coronary Risk Factor Intervention Project (STRIP) [3] and from the ages of 8–10 yr throughout adolescence studied in the Dietary Intervention Study in Children (DISC) [4]. However, efficacy and safety of low fat diet is based on epidemiological reports and insulin resistance, endotelial function in adolescents are factors to interfere with lipid lowering.

Current View of Nutraceuticals and Bioactive Foods in Lipid Lowering

The authors believe that lipid lowering effect based on the mixed reports of verbal testimonies or epidemiological surveys is insufficient claim in favor of lipid lowering by foods. However, scientific evidences exist in favor. The use of margarines (3 servings daily) high in either plant stanol esters or plant sterol esters can reduce LDL-C 10–15% when added to a low-fat diet. Water-soluble fibers such as psyllium may lowering of 5–10% LDL-C. Soy protein lowers VLDL-C and TG and increases HDL-C without effect on LDL-C. Supplementation of a low-fat diet with an ω-3 fatty acid (docosahexaenoic acid 1.2 g/d) increased the large sized LDL 91% and reduced the small sized LDL 48% [3]. However, the prescription of ω-3 fatty acids is not yet approved by the FDA. Garlic does not lower LDL-C in hyperlipidemic children [3]. Niacin is choice in FH homozygotes (55–87 mg/kg·d in divided doses), due to the significant reduction of VLDL and LDL production (Figure 1). Use of a fibrate (48, 96, or 145 mg/d) is limited to that adolescent with TG over 500 mg/dl, who may be at increased risk of pancreatitis. Fish oils (1–2 g/d) lower TG by decreasing TG biosynthesis (Figure 2). A recent report on bioactive foods reviewed the significance of plant sterols, fibers, flavones, anioxidants and polyunsaturated fatty acids in lipid lowering with biochmical basis of cardiovascular prevention [21]. Overall, a diet low in fat in children with dyslipidemias appears safe and efficacious when performed under supervision. Medical and nutritional support is necessary to reinforce good dietary behaviors and ensure nutritional adequacy.

Drug Therapy in Lipid Lowering

The drug therapy is aimed to lower significantly elevated LDL-C during Tanner stage II in males and post-menstrual stage in girls if the LDL-C is > 190 mg/dl and no history of CVD; or LDL-C > 160 mg/dl with family history of premature CVD or the metabolic syndrome is present. In this direction, two major clases of effective drugs, stains and bile acid sequestrants (BAS) have surfaced to combat with elevated LDL-C (Figure 1). The main mechanism of lipid lowering is the release of the sterol regulatory element binding protein (SREBP) from the cytoplasm into the nucleus, where SREBP binds to the SRE of the promoter of the LDL receptor gene and increases the number of LDL receptor, to decrease LDL-C. SREBP also up-regulates the gene for hydroxymethylglutaryl coenzyme A (HMG-CoA) reductase to inhibit this enzyme.

- *HMG-CoA reductase inhibitors:* Statins reduce hepatic cholesterol by inhibiting HMG-CoA reductase and decreasing cholesterol biosynthesis. It improves endotelial function. Atorvastatin, lovastatin, pravastatin, and simvastatin are approved by the FDA for use in adolescents with FH.
- *Bile Acid Sequesterants:* BAS are safe drug recommended by NCEP for lipid-lowering. Cholestipol, Colesevelam (625-mg tablets, three or six per day), Cholestyramine drugs reduce lipids by 15%.
- *Cholesterol absorption inhibitors:* Ezetimibe, blocks the absorption of cholesterol and plant sterols and reduce hepatic cholesterol (Figure 2).

2.4. Modern View on Dyslipidemia Management and Nutrition Therapy

In the modern world, fatty fish, fish oil, walnuts, oatmeal, and oat bran, and foods fortified with plant sterols or stanols are advocated to help control cholesterol. Some studies have shown that a diet combining these "superfoods" may work to reduce your "bad cholesterol" or LDL cholesterol levels in blood. ADA advocated a golden rule: Plant sterols reduce dietary cholesterol absorption; Nuts kill LDL-cholesterol; Omega-3 fatty acid rich fish (Salmon, tuna, sardine) bring down triglycerides and total cholesterol; Oat meal brings down LDL-cholesterol; Soy protein brings down LDL-cholesterol and triglycerides without lowering HDL cholesterol; Nuts increase HDL-cholesterol.

National guidelines indicate that patients with elevated LDL cholesterol should consume less than 7% calories from saturated fat and less than 200 mg cholesterol means limited trans fatty acids and supplementation of monounsaturated omega 3 fatty acids and stanol rich margarine, soy products, cereals with vegetables. Major goal is specific to lowering LDL cholesterol, raising HDL cholesterol, and limiting serum triglycerides (representative of very low density lipoprotein [VLDL]) to control dyslipidemia and avoid atherogenesis and diabetes.

Table 5. Food supplements in dyslipidemia management and control. Source: Third Adult Treatment Panel of National Cholesterol Education Program 2002 [19]

Bioactive foods	Nutrition contents	Readymade recipes
Plant sterols and stanols (2 grams a day)	Grains, vegetables, fruits, legumes, nuts, and seeds.	Margarine spreads, orange juice, cereals, granola bars, some cooking oils, salad dressings, milk, yogurt, snack bars, and juices
Nuts (A handful daily intake)	Walnuts, Almonds rich in protein, fiber, healthy monounsaturated fats, vitamins, nutrients, and antioxidants	salty, fatty, and high calorie snack bowls of dingy, smoky bars
salmon, tuna, sardines, herring, mackerel, and trout (3 ounces a day)	omega-3 fatty acids	fish fry, fish chips, fish oils
Oatmeal fiber or bran (25 grams of fiber a day)	soluble fibers	boiled pudding cup
Soyabean (50 g per day)	Soy meal, a side dish, a snack, or drink	tofu, soy nuts, soy burger, soy cheese, edamame, tempeh, miso, soy flour

Mechanisms

On a molecular level, plant sterols and stanols look a lot like cholesterol. So when they travel through your digestive tract, they get in the way. Sterols or stenols can prevent real cholesterol from being absorbed into your bloodstream. Instead of clogging up your arteries, the "bad" cholesterol just goes out with the other waste as shown in Figure 2. Metabolic mechanistic basis of dyslipidemia disorders can be described at various end points including LDL receptor activity [22], Familial hypercholesterolemia either heterozygots or homozygots, inability of ApoB-100 binding with LDL-R [23], autosomal recessive hypercholesterolemia (ARH) [24], mutations in proprotein convertase subtilisin-like kexin type 9 (PCSK9) [25]. Basically overproduction of VLDL and LDL is the key of FCHL, hyper-ApoB, LDL subclass pattern B, familial dyslipidemic hypertension, and syndrome X of Reaven at genetic level [25]. Briefly, abnormal free fatty acid (FFA) metabolism in FCHL and other small, dense LDL syndromes may reflect the primary defect in these patients or impaired insulin-mediated suppression of hormone-sensitive lipase in adipocytes leads to an elevation in FFA [26, 27]. Elevated FFA may drive hepatic overproduction of TG and ApoB, leading to a two- to threefold increased production of VLDL and the dyslipidemic triad. Insulin resistance also interferes with normal upregulation of lipoprotein lipase by insulin, leading to decreased lipolysis of TG in VLDL, as well as in intestinally derived TG-rich lipoproteins.

Table 6. Web sites helpful for patient education and information

Web address	Materials available
http://www.bellinstitute.com/nutrition/hn/nem.htm	Behavior change focused based on the ATP III with a focus on soluble fiber from oat cereal. Brochure—Destination: Healthy Heart brochure.
http://www.ndep.nih.gov	National Diabetes Education Program Brochure—Be Smart about your Heart: Control the ABCs of Diabetes
http://www.nhlbi.nih.gov/chd/lifestyles.htm	Therapeutic Lifestyle Changes information for consumers
http://www.deliciousdecisions.org	Interactive AHA materials including advice on cooking and shopping
http://www.nhlbi.nih.gov/about/oei/index.htm	Obesity Education Initiative; weight control information for consumers and health professionals
http://www.soyfoods.com	Information on soy products and amounts of soy protein and recipes

AHA—American Heart Association; ATP III—Third Adult Treatment Panel of the National Cholesterol Education Program.

Table 5. Approximate LDL cholesterol reduction achievable by dietary modification

Dietary component	Dietary change	Approximate LDL reduction, %
Saturated fat	< 7% of calories	8–10
Dietary cholesterol	< 200 mg/d	3–5
Weight reduction	Lose 10 lb	5–8
Soluble fiber	5–10 g/d	3–5
Plant sterol or stanol esters	2 g/d	6–15
Cumulative estimate	—	20–30

LDL—low-density lipoprotein.
(*Adapted from* the National Cholesterol Education Program, Third Adult Treatment Panel; http://www.nhlbi.nih.gov/guidelines/cholesterol/atp3_rpt.htm [V-32].)

This paradigm may also result from a defect in the normal effect of acylation stimulatory protein which is to stimulate the normal incorporation of FFA into TG in the adipocyte [28].

Medical Nutrition Therapy to Lower LDL Cholesterol

Diet limited in the LDL-raising nutrients such as saturated fatty acids, cholesterol, and trans-unsaturated fatty acids were suggested by NCEP's ATP III and the ADA's nutrition principles for the benefit of low dietary saturated fat to maintain a serum LDL cholesterol below 100 mg/dL in diabetes [29, 30]. We describe following advice for each diet supplement in following sections based on NCEP, ADA recommendations.

Saturated Fatty Acids

Dietary saturated fatty acids decrease synthesis and activity of LDL receptors, providing elevated serum LDL cholesterol conducive to atherogenesis [31]. Every 1% increase in calories from saturated fat increases serum LDL cholesterol by approximately 2% [5•]. ATP III guidelines limit this saturated fat to less than 7% of calories. A meta-analysis concluded the previous NCEP Step 1 diet (30% calories from total fat, less than 10% calories from saturated fat and 300 mg of cholesterol) lowered LDL by 12%, but the further limitations of the Step 2 diet (7% calories from saturated fat, 200 mg of cholesterol) lowered LDL an average of 16% [32]. *Diet advice:* Keep lower saturated fat to less than 7% of calories; select animal products low in fat, such as skim or 1% fat milk and other low-fat dairy products; daily meat intake of lean beef, poultry, or fish intake less than 6 oz [33]; less saturated vegetable fats (palm and coconut oils) (see Table 4).

Trans Fatty Acids

In American diet, 2.6% calories comes from trans fatty acids and 11% calories comes from saturated fat [29]. Trans-unsaturated fatty acids are formed from polyunsaturated fats, such as corn and soybean oil upon hydrogenation to produce stick margarine and fats with greater shelf-life. These also elevate LDL cholesterol, and LDL cholesterol. ω-3 fatty acid (docosahexaenoic acid 1.2 g/d) supplemented low-fat diet did not lower LDL-C but significantly increased the largest LDL subclass 91% and decreased the smallest LDL subclass 48% [34]. Metabolic diets containing various fat sources (*eg,* butter, soybean oil, semi-liquid margarine, shortening, or stick margarine) elevate LDL cholesterol levels while reduced LDL by 5% lower on stick margarine, 9% lower on tub margarine, 12% lower on soybean oil, and 11% lower on margarine [35, 36]. *Diet advice:* Less saturated fat and trans fatty acids- polyunsaturated and monounsaturated fat intake (7% in tub margarine, and less than 1% in semi-liquid margarine); cookies, crackers, and doughnuts, french fries or chicken) [37].

Dietary Cholesterol

Cholesterol in diet increases total serum cholesterol (10 mg/dL for every 100 mg per dietary cholesterol per 1000 calories) [29]. Best example is limited egg yolk intake (American Heart Association (AHA) recommends one yolk per day and ATP III recommends two per week) to keep dietary cholesterol less than 200 mg/d [35].

Monounsaturated Fatty Acids

Monounsaturated cis fatty acids in olive oil, canola oil, avocado oil, pecan oil, and peanut oil lowers LDL and triglyceride [29]. Low-calorie diet with monounsaturated fat lowered LDL, optimum LDL or HDL in type 2 diabetes [38]. *Diet advice:* Olive or canola oil and salad dressing with avocado on sandwiches, snack on pecans, or pack a peanut butter sandwich for lunch.

Plant Sterol and Stanols

The ATP III recommends adding more than 500 mg/d plant stanol or sterols such as margarine 2 g/d to the diet to promote additional LDL lowering reduce absorption of endogenous and exogenous cholesterol (three doses lowers LDL cholesterol by 10%) [29, 39, 40, 41]. However, stanol or sterols pose concern of potential addition of calories. For example, high stanol or sterol esters reduce absorption of fat-soluble vitamins and carotenoids and it requires more fruit and vegetable intake [42, 43]. *Diet advice:* 2 g of sterol or stanol per day in two servings of margarines: Benecol (McNeil Consumer Products, Fort Washington, PA) or of sterol in 'Take Control' (Unilever, Englewood Cliffs, NJ). These margarines are available in both regular and light versions.

Wild Foods

Wild foods are mainly wild plants, wild vegetables, wild fruits and wild mushrooms. Mushrooms, flowers, fruits, wine berries were reported beneficial in cardioprotection by

He et al. 2007 [44], Heidemann et al. [45], INTERHEART study [46]. Other epidemiological studies brought out other factors BMI, waist circumference, alcohol consumption, low physical activity, hypertension, diabetes, C-reactive protein, Fetuin-A and insulin resistance as risks not protected by wild foods. For full description on value of Mediterranean soup (mixture of tomato, graps, raisins, carrot, spinach, walnuts, almonds, lin/chia seeds, olive oil) in acute coronary syndrome (ACS) patients with pro-inflammatory effects and challenges, readers can read the web document written by first author (http://www.scribd.com/doc/22527813/Wild-Foods-in-Cardioprotection) highlighting the 'Columbus Concept' originally advocated by DeMeester [47].

Soluble Fiber

ATPIII recommended the inclusion of 10 to 25 g of soluble fiber such as grain cereal, oat meal and 5 g psyllium daily for 8 weeks supplemented vitamins and minerals in the diet lowers 7% LDL cholesterol, 10% plasma glucose and serum triglycerides in men with type 2 diabetes and hypercholesterolemia on NCEP Step 1 diet [48]. *Diet advice:* fruits, vegetables and grains high in soluble fiber 10 to 20 g /d (lima beans, brussel sprouts, citrus fruits, papaya, raisins, winter squash, and oats) supplemented with psyllium-containing Metamucil (Proctor and Gamble, Cincinnati, OH) or psyllium-fortified foods [29, 42].

Soy

Soy food has high isoflavones, protein and low saturated cholesterol fat. Soy lowered LDL cholesterol by 4% to 8% with cardiovascular benefit as recommended by both US Food and Drug Administration and AHA [49, 50].

For patients with diabetes, glucose control and insulin resistance are also improved by soy supplementation [51]. Major benefits were; lowering of VLDL-C, TG, plasma homocysteine; increased HDL-C, lipoprotein(a) and no change in LDL-C and hemoglobin A1c [52, 53, 54]. Overall, role of individual soy components is unclear. Soy appears to have cardiovascular benefit, likely through multiple mechanisms. *Diet advice:* Tofu, tempeh, and miso, soy nuts, soymilk, soy yogurt, and soy cheese [55].

Macronutrient Distribution

Diabetic patients with dyslipidemia are sensitive to the intake proportion of carbohydrate to fat. ADA recommendations up to 60% to 70% of calories from combined intake of monounsaturated fat and carbohydrate [56]. Practice to decrease carbohydrate intake is by increased intake of monounsaturated fat. Interesting point is that Low-fat diets may raise serum triglycerides while high carbohydrate intake and adequate soluble fiber may moderate any detrimental effect of a lower-fat intake. The low-fat (10% fat, 4% monounsaturated fat) and the monounsaturated-fat diet (32% fat, 7% saturated) lowered LDL 10% and 17%, respectively [38].

Combined Effects

Many nutritional strategies can lower LDL cholesterol. ATP III estimated that the combination of its major dietary principles could lower LDL cholesterol 20% to 30% (Table 2) [29]. A more global approach to combined nutrition treatment strategies with type 2 diabetes and microalbuminuria suggested diet, exercise, and vitamin supplementation combined with medication andquarterly visits with a team of physician, nurse, and dietitian with result of better lipid profiles, better glycemic control, and good cardiovascular protection [57].

Clinical Trials to Modify Residual Cardiovascular Risk beyond LDL Cholesterol Lowering

As discussed before, it is crucial to modify all the atherogenic risk factors for better outcomes in patients during dyslipidemia or later with atherosclerotic vascular disease. In following section, we describe the new options of nutraceuticals alone or in combinatorial way with statins in differtent trials. However, stains have come up as potential drugs to control lipids but with serious side effects and difficult compliance [58]. To modify the risk beyond statin therapy, the following combined nutrition therapies with drug schedules are available to manage dyslipidemia in order to avoid risk of atherosclerosis or metabolic syndrome with diabetes.

1. Omega-3 fatty acids: JELIS, COMBOS, AFFORD
2. Niacin: CDP, SEACOAST, OCEAN, ARBITER-1,2,3
3. Fibrates: FIELD, HHS, VA-HIT,
4. Combination of statins with niacin
5. Combination of statins with fibrates.

Supplementation of Omega-3 Fatty Acids

The 2007 National Lipid Associations (NLA) safety task force concluded that omega therapy is a safe therapeutic option for lowering TG. [59] Observational studies have shown several cardiovascular benefits such as a decrease in cardiac dysrrythmias, sudden cardiac death, and a decrease in blood pressure. [60] The mechanism of action of omega-3 fatty acids in the reduction of TG is unclear. There is evidence that omega-3 fatty acids increase TG clearance from circulating VLDL particles by increasing lipoprotein lipase (LPL) activity. Some studies have shown an increase of HDL-C with high doses of omega-3 fatty acids. In the JELIS study (Japan Epa Lipid Intervention Study), a combination of omega-3 fatty acids and statin was compared with statin monotherapy. There was a 19% reduction in major coronary events by the combination therapy as compared to statin alone. [61] Another trial, COMBOS (COMBination of prescription Omega-3 plus Simvastatin) which also showed that a combination of omega-3 fatty acids and simvastatin reduced non-HDL-C, TG and raised HDL as compared to statin monotherapy. [62] The AFFORD trial (Atorvastatin Factorial with Omega-3 fatty acids Risk Reduction in Diabetes) [63] did not show any benefit of residual cardiovascular risk reduction in the diabetic population. However, dietary supplement of omega-3 fatty acids is not subject to FDA regulation and thus higher doses of fish oil supplement may be required to be equivalent to the prescription form of omega-3 fatty acids (Lovaga). [59] Omega-3 fatty acids may reduce the synthesis of TG and VLDL and increased APO-B degradation (transcription of ATP binding protein transporter) to increase HDL-C to facilitate HDL-C mediated cholesterol efflux from peripheral hepatic cells.

Niacin

Niacin has long been recognized for its lipid-modifying effect. It is advocated as ideal in combinatorial nutrition supplement. It has a well established safety profile based on clinical evidence over 20 years. It raises HDL-C and had been used for past 5 decades. [64] The Coronary Drug Project (CDP) randomized trial conducted between 1966 and 1975 with a mean follow up of 6.2 years in men with history of previous MI showed the following: 1. primary end point of mortality did not decrease in the niacin group; 2. significant reduction in composite outcome of CHD death, non-fatal MI, and cerebrovascular events occurred; 3. significant reduction of cardiovascular surgery (47%) in the niacin group; 4. After 9 years of trial termination, there was an 11% (P <0.0004) mortality reduction in the niacin group [65] Biochemically, niacin reduces hepatic synthesis of TG, VLDL particles, and catabolism of APO A while it increases APO A-1, HDL-C and degradation of APO-B [66, 67]. Niacin may also enhance ABC A-1(ATP binding cassette) transporter transcription leading to HDL-mediated cholesterol efflux [68]. Combination therapy with niacin and statins has shown greater efficacy and safety in patients with dyslipidemia evaluated in SEACOAST (Safety and Efficacy of a combination of Extended Release Niacin and Simvastatin Trial) and OCEAN (Open label Evaluation of the safety and Efficacy of a combination of Niacin ER and Simvastatin) trial [69, 70]. In SEACOAST-I trial, fixed dose combination of niacin ER and simvastatin (1000/20 and 2000/20) showed a significant dose-related improvements in non-HDL-C, HDL, TG, and lipoprotein(a) compared to simvastatin monotherapy. The most notable results were 24% increase in HDL-C, 38% reduction in TG, and 25% reduction of lipoprotein(a) in a group treated with niacin/simvastatin combination [69]. SEACOAST-II trial showed a 17.1% reduction in primary end point of non-HDL-C with niacin/simvastatin 2000/40 compared to a 10.1% reduction in simvastatin 80 mg monotherapy. Randomized OCEAN trial open label, multicenter study evaluated the safety and efficacy of a fixed dose combination of niacin and simvastatin in patients with elevated non-HDL [70]. The primary end point was long-term safety and secondary endpoints were the serum levels of non-HDL-C, IDL-C, and TG. A subgroup of patients failed to reach their goals with simvastatin as baseline therapy. However, 82% achieved non-HDL goals, 85% reached HDL goals, 67% reached HDL goals, and 64% reached TG target (65% reached all combined goals). Other trial HDL-Atherosclerosis Treatment Study (HATS) was a double-blind, placebo-controlled trial in which 160 patients with low HDL and low LDL < 145 mg/dL were enrolled for coronary angiography at baseline and 2-year follow up with simvastatin with niacin and the endpoints were the angiographic change in CAD and occurrence of first cardiovascular event such as MI, death, coronary revascularization, and stroke. Patients showed a slight progression of angiographic CAD in the placebo group (3.9% changes) and regression of angiographic CAD (0.4% change) in the treatment group with niacin and simvastatin. Investigators concluded that niacin/simvastatin therapy in CAD patients with low HDL resulted in slight regression of atherosclerosis over a 3-year period. [71] Arterial Biology for the Investigation of the Treatment Effects of Reducing cholesterol (ARBITER) a double-blind, placebo-controlled study of once-daily niacin ER 1000 mg added to statin therapy in CAD patients with low HDL-C levels showed primary end point as change in carotid intima-media thickness (CIMT) after 12 months (0.044 mm in the placebo group compared to 0.023 mm in the niacin group). In other progressive ARBITER-2 for blinded 12 months study and ARBITER-3 for an additional 12 months showed end point as significant regression of atherosclerosis measured by CIMT at both 12 months and 24 months compared to statin therapy alone. [72] However, niacin intake is not risk free from common side effect, flushing as a marker of high lipid response to niacin therapy. [73]

Fibrates and Fenofibrates

Several studies have shown the cardiovascular benefits of fibrate therapy. The Helsinki Heart Study (HHS)42 showed a 71% reduction in CHD in patients taking gemfibrozil compared to patients receiving placebo, and similar results were seen in VA-HIT (Veterans Affairs-High density lipoprotein Intervention Trial) which showed a 41% reduction in CHD and stroke with gemfibrozil compared to placebo in a subgroup of patients with DM [74]. The BIP trial (Bezafibrate Infarction Prevention) evaluated the long-term cardiovascular benefit of bezafibrate therapy [75]. This study demonstrated significant long-term cardiovascular protection which was attenuated by unbalanced use of non-study lipid-lowering drugs. Fibrates work by activating peroxisome proliferatoractivated (PPRA) alpha receptors which modulate several aspects of lipid metabolism by increasing expression of APO A-1, APO-11, and ABCA1, and lowering expression of APO C-111. They also increase HDL-C particles. Fibrates increase LPL synthesis which clears VLDL clearance and lowers TG. They also increase B-oxidation of fatty acids leading to a decrease in TG and VLDL production. [76, 77] Fenofibrate therapy was tested in FIELD (Fenofibrate Intervention and Event Lowering in Diabetes) trial. [78] This study was planned to extend the findings of the HHS and VA-HIT studies by investigating the long-term effect of fenofibrates in the largest trial of patients (total 9795) with type 2 DM with a 5-year follow up. The fenofibrates showed a 11% reduction (insignificant) in primary end point of CHD but the study did show a 24% reduction in non-fatal MI and a 21% reduction in coronary revascularization (significant). The interesting aspects of this study were certain prespecified tertiary endpoints like effects on microvascular complication of DM such as microalbuminuria, diabetic retinopathy, and amputation due to non-traumatic causes. In the overall analysis for prespecified endpoints, there was a regression of microalbuminuria in the fenofibrate group. [79] There was also a beneficial effect in the subgroup with diabetic retinopathy. The fenofibrate group showed a 31% reduction in the need for first laser treatment compared to placebo, and benefits progressively increased thereafter.47 Fenofibrates also reduced the number needing non-traumatic amputation in these diabetes patients. This demonstrates the beneficial effects of fenofibrate in preventing macro- and microvascular complications of DM.

As fibrates modify all aspects of dyslipidemia, their use in combination with statins is very attractive. [80] Both agents have the potential for myopathy and the risk of adverse events depends on the pharmacokinetic interaction between their affects on statin metabolism and clearance.49–53 Several studies have shown that gemfibrozil interferes with the metabolism of statins by inhibiting glucoronidation. This possibly can raise statin levels, predisposing the patients to myopathy. [80,81] In contrast, fenofibrate does not interfere with statin metabolism and therefore may be safer to use in combination with statin therapy [81].

Because of this pharmacokinetic interaction, the National Lipid association (NLA) safety task force has recommended avoidance of usage of gemfibrozil in combination with statins, and fenofibrates may be the preferred fibrate to use in combination with statins [83].

It has been also stated in NCEP ATP-III update in 2004 that, unlike gemfibrozil, fenofibrates does not increase rate of myositis when used in conjunction with moderate doses of statins.55 There are several ongoing trials to address this issue of combination therapy of omega-3 fatty acids, niacin, fibrates, and statins [84, 85].

Less Known Facts on Indian Tribals and Dyslipidemia

The Indian Tribal: Kurichias as Example

The Kurichias tribal of Wayanad plateau, inhabits at the trijunction of Kerala, Karnataka and Tamilnadu in South India. Elderly Kurichias population is reported risk free from dyslipidemia or other diseases while survive on poor food availability (9,10). It is important to understand their exercise filled life style how it affects their health. They live in enormous joint families and follow a commune type of family organization. The following description is extracted from the first time report on Kurichias by eminent nutritionist R.B.Singh et al. [18]. In his words, "The elderly population of Kurichias constitute about 12% in their general population which is two fold greater compared to elderly subjects (6.5%) among other Indian tribal populations. The proportion of elderly Kurichias is almost closer to that of many developed countries (10-14%) in Europe and West. Kurichias live in scattered homesteads on leveled grounds adjacent to their agricultural fields, and the distance between one such settlement to another may range from 2 km to 10 km. Their massive joint families maintain a matrilineal heritage, which can be understood in several aspects such as leadership and its succession, property inheritance, management of the family on an egalitarian basis. The rural and land owning Kurichias strictly adhere to their culture following matriline age. The older generation also continues to have the leadership control over the family. The senior most of the family member is given due importance in all walks of life, even after retiring from head-manship, The excellent ecology is another factor for the longevity of the Kurichias inhabiting Wayanad region of Kerala receives maximum rainfall in the State of Kerala. Kurichias possess strong physique although they have recently started farming in which all the members of the family participate in cultivation. They are fond of hunting and the work environment of Kurichias is located in the midst of two hills. Kurichias daily walk for 10-20 miles every day which is characteristic of hunting and farming as hunter-gatherers. Obesity and overweight are not seen among [16, 17]. Ownership of land is collective among the Kurichias, which facilitates them to lead a inexpensive life. They are not interested in the accumulation of wealth but now they have started storing food for next few months. Elderly people exhibit anxiety over the ambitious attitude of younger generation who want to accumulate wealth. Causes of deaths among Kurchias are mainly fall, bites, drowning and killing during hunting or other natural calamities". Cancer, heart attack, hypertension, obesity and diabetes mellitus are uncommon among Kurichia hunter-gatherers [16, 17].

Food Intakes among Indian Hunter-Gatherers

Long life expectancy among Indian hunter-gatherers has attracted attention to explore the food consumption pattern of the Indian hunter-gatherers [8, 9]; vegetables, leaves, nuts, tubers, fruits, whole grains, eggs, fish, honey. Interestingly, Indian Kurichia hunter-gatherers are non-vegetarians. R.B. Singh et al. [18] reports a match between ancient diet habit with Kurichias diet as "Meat is taken mostly from the untamed animals and the time of hunting was possibly, early in the morning, at sun rise when animals come for drinking of water at the river banks. They are fond of taking fresh meat that is hunted by them selves for consumption. Since the government of India has imposed ban on hunting in the forest, their meat intake is now drastically reduced as given in the Tables 5-7. Dietary intakes assessed by 3-days dietary recall, were recorded in a questionnaire, by the dietitian and by asking food intake frequency in the last one month. Food models, food measures and food portions were showed to individuals with the help of an educated Kurichia, who knew both the languages, during dietary assessments.

Table 7. Food consumption pattern among Indian hunter-gatherers

Morning: Most take milk or nothing but few prefer to have black tea (Fresh tea leaves that is available in their paddy fields will be taken, boiled in the water and then take with sugar)
Breakfast: Majority of the people (around 95%) prefer to have Gangi in the morning for about 200 g to 300 g. (Gangi: Brown rice mixed with Ragi and then kept over night in the water for fermentation and then take the semi solid mixed with water). This will be supplemented with some self made pickle.
Lunch: They prefer to have brown rice only (cultivateded from their fields without use of fertilizers). Rice cooked (200-300g) will be supplemented with green leaves (100-200g,seasonally available) and a fried fish,100-200g (fresh water) available in their canals. They take some quantity of seasonal fruits, 100-200g daily.
Evening: Few people prefer to have black tea (Fresh tea leaves that is available in their paddy fields will be taken, boiled in the water and then take with sugar)
Night: They prefer to have brown rice only (yielded from their fields only). Rice will be supplemented with seasional green leaves (100-200g/day) and a fried fish(150-300day) available in their canals and buttermilk (100-300ml) They take some quantity of fruit (100-200g)

The dishes taken by Kurichias will predominantly occupy with green leaves like Cassia tora, Alternanthera sessilis etc, vegetable like hydrocotyle, clerodendron, nymphaea nouchali and roots like ceropegia and elaeocarpus [16, 17].

Sugar is a recent addition during the last 50 years. The main foods are; brown rice, seasional green leaves and fruits, roots and tubers, fish, and butter milk.

Alcohol and tobacco intakes are considered evils which are uncommon. These foods are rich sources of w-3 fatty acids, antioxidants, essential amino acids, vitamins and minerals and soluble fiber and are low in saturated fat, w-6 fat and trans fat. Sugar is the only refined carbohydrate.

The food and nutrient intakes among Kurichias appear to have similarity with the food advice in the ancient scriptures classically known as Rajasik diet (see Table 3)".

Table 8. Food and nutrient intake among hunter-gatherer and Western population

Food and nutrient	Hunter-gatherer	Western population	Asians
Energy density	Low	High	Low
Protein	High	Low-moderate	Low
Animal	High	Low –moderate	Low
Vegetable	Very low	Low –moderate	Low
Carbohydrate	Low-moderate(slowly absorbed)	Moderate-rapidly absorbed	High-Slow
Fiber	High (>30g)	Low (<15g)	High
Fat	Low	High	Low
Animal	Low	High	Low
Vegetable	Very low	Low (0.2g/day)	High
Total w-3	High(2.3g/day)	High 15-20	0.5-0.85g
Ratio w-6:w-3	Low 2.4	low	25-50
Vitamins and minerals	high		High

Modified from Eaton et al 1998 [11]and Singh et al 2011 [18].

Table 9. Estimated fatty acid consumption in the late Paleolithic period

Sources	Fatty acids(g/day) en 35.65/day
Plants	
Linoleic acid	4.28
Alpha-linoleic acid	11.40
Animal	
Linoleic acids	4.56
Alpha-linolenic acid	1.21
Total	
Linoleic acid	8.84
Alpha linolenic acid	12.60
Animal	
Arachidonic acid(w-6) (AA)	1.81
Long chain w-3 fatty acids	
Eicosapentaenoic acid(w-3)(EPA)	0.39
Docosatetraenoic acid(w-6) (DTA)	0.12
Docosapentaenoic acid(w-3)(DPA)	0.42
Docosahexaenoic acid(w-3)(DHA)	0.27
Total long chain w-3 fatty acids	1.20
Ratios of w-6/w-3	
Linoleic acid/alpha linolenic acid+	0.70
AA+DTA/EPA+DPA+DHA	1.79
Total w-6/w-3	0.77

Modified from Eaton et al [11] and Singh et al 2011. [18].

Table 10. Nutrient composition in the late Paleolithic and current recommendations

Nutrient per day	Late Paleolithic	Current recommendation
Total dietary energy%		
Protein	33	12
Carbohydrate	46	58
Fat	21	30
Alcohol	-0	moderate alcohol
P/S ratio	1.41	1.00
Cholesterol, mg	520	300
Fiber, g	100-150	30-60
Sodium,mg	690	1100-3300
Calcium, mg	1500-2000	800-1600
Ascorbic acid, mg	440	60

Modified from Eaton et al [11] and Singh et al 2011. [18].

Table 11. Ethnic differences in fatty acid levels in thrombocytes phospholipids and percentage of all deaths from cardiovascular disease

	Europe and USA %	Japan %	Greenland Eskimos %
Arachidonic acid(20:4w6)	26	21	8.3
Eicosapentaenoic acid(20:5w-3)	0.5	1.6	8.0
Ratio of w-6/w-3	50	12	1
Mortality from cardiovascular disease	45	12	7

Modified from Singh et al 2011. [18].

Table 12. Fatty acids ratio in the diets

Subjects	w-6/w-3	
Paleolithic	0.79	Estimated
Greece prior to 1960	1.00-2.00	Current 7.10
Japan	4.00	Early 1-2
India, rural	5-6.1	Prior to 1960, 3-4
India urban	38-50	Prior to 1960, 5-10
UK	15.00	Prior to 1960,10.00
Northern Europe	15.00	Prior to 1960,10.00
USA	16.74	Prior to 1950 7-8
Eastern Europe	20-25	Estimated
Indian hunter-gatherers	1.00-2.00	Estimated

Modified from Singh et al 2011 [18].

3. Archeological Concepts of Nutrition and Health

A fundamental view of evolution in relation to health promotion has been seeded that the genetic makeup of contemporary humans shows very minor difference from that of the modern humans who appeared in Africa between 100,000 and 50,000 years ago [11, 12]. The human evolution has been comparatively rapid during the past 50,000 years, as revealed by the molecular geneticists. There have been marked changes in the food supply with the development of agriculture about 10,000 years ago from now. Initially, it was related with hunting and carnival habits. However, only non-significant change in our genes occurred during the past 10 centuries, due to presence of w-3 fatty acids, essential amino acids, antioxidants, vitamins and minerals in the human prepared diet and awareness of agriculture. Conceptually, spontaneous mutation rate for nuclear DNA is estimated at 0.5% nucleotides changed per million years. At this rate, over the past 10,000 years there has been small time for very little change in our genes, possibly 0.005%. Our genes appear to be similar to the genes of our ancestors during the Paleolithic period 40,000 years ago, the time when human genetic profile was finally evolved as established profile. Man *homosapian species* appears to live in a developed nutritional environment which completely differs from that environment for which originally our genetic constitution was selected and evolved. However, during the last 100-160 years, daily dietary intakes have changed significantly, causing increased intake of saturated fatty acids (SFA), trans fatty acids and linoleic acid and decrease in w-3 fatty acids, obtained from grain fed cattle tamed at farm houses, rather than meat from running animals. In general, modernization and nutrition industries have made daily increase in the intake of refined carbohydrates and decrease in the intake of complex carbohydrates, essential amino acids, minerals, vitamins and antioxidants. For interested readers, food and nutrient intake among hunter-gatherers now and during Paleolithic period are given in the Tables 1-5. Clearly, comparison shows a marked reduction in consumption of w-3 fatty acids, antioxidants, vitamins and minerals and proteins and significant increase in the intakes of carbohydrates, (mainly refined,), fat (saturated, trans fat, linoleic acid) and increased salt compared to Paleolithic period [11-17]. We believe and propagate our idea that "these said dietary changes in conjunction with present time sedentary life style with mental strain, pollution, tobacco consumption and alcoholism, particularly after year 1910, during the last 100 years might have caused a biochemical remodulation or damage to our genes, in order to combat the physical, environmental conditions resulting with the emergence of non-communicable internal organ diseases". Recently, similar views on protective effects and dietary practices of Paleolithic diets was highlighted in the form of Mediterranean diet, Indo-Mediterranean diet, Japanese diet, DASH diet documented in several studies [86-95]. The ancient concepts of nutrition, particularly Rajasic diet (Sedentary life style in modern times) and the different foods consumed by Indian tribal hunter-gatherers, support the importance of natural foods in diet and consumption of less refined food similar to diets consumed by Kurichias tribals. These studies have clearly demonstrated the beneficial effects of these natural foods in diets, against coronary artery disease, hypertension, type 2 diabetes mellitus and cancers [17, 86-95].

4. Dietary Fat Intake and Fatty Acid Ratio

Recently 'fatty acid ratio' was highlighted based on unsaturated and saturated fats in the diet and overall distribution to explain good health or risk of atherogenesis after oxidative stress. We describe 'fatty acid ratio' in this section and oxidative stress in next section.

The fatty acid ratio in 'Columbus concept' of diet means that humans evolved on a diet that was low in saturated fat and the amount of w-3 and w-6 fatty acids was quite equal. [14, 15] On the lines of 'Columbus concept' "Nature recommends to ingest fatty acids in a balanced ratio(poly-unsaturated:saturated=w-6:w-3=1:1)as part of dietary lipid pattern in which monounsaturated fatty acids (M) is the major fat(P:M:S=1:6:1). These ratios represent the overall distribution of fats in a natural untamed environment". (www.columbus-concept.com) [14, 15].

According to the 'Columbus concept' ancient foods included egg, milk, meat, oil, whole grain foods, all rich in w-3 fatty acids, similar to wild foods, consumed about 160 years ago from now. Furthermore, blood lipid composition reflects one's health status predicted by: (a) circulating serum lipoproteins and their (LDL+VLDL)/Total Cholesterol ratio provides information on their atherogenicity to blood vessels; and (b) circulating plasma fatty acids, such as w-6/w-3 fatty acid ratio, gives indication on proinflammatory status of blood vessels. Both factors (a) and (b) are phenotype-related and both depend on several genetic, environmental and developmental factors. Hence, they appear as universal markers reflecting physical, mental, social and spiritual health that are more clearly described in ancient scriptures of Indian Ayurved 'Charak Samhita' [96] and Hebreu "Tsim Tsoum concept" (http://www.tsimtsoum.net/editorials/tsimtsoum_editorial_2009-Kosice-14th-WCCN-and-5th-ICCD.pdf) . The word 'Tsim Tsoum' is derived from Hebreu and it is similar to 'ying yang' in Chinese and is the indicator of total health. Blood cholesterol may be central in these approaches, however, it also includes simultaneous approach of protecting mind-body interactions against environmental factors. Author has described in detail to emphasize other contributory factors to influence mind-body interactions and life style. [http://www.scribd.com/doc/19575964/Traditional-Therapies-Lessons) [96].

8. Oxidative Stress in Atherogenesis due to Dyslipidemia

Increasing numbers of studies have demonstrated that oxidative stress, i.e. dysregulation of cellular redox state, plays a pivotal role in the pathogenesis of atherosclerosis, especially vascular endothelial dysfunction [97]. Superoxide anion is formed by univalent reduction of molecular oxygen. Although several enzymes are involved in the generation of superoxide anion, including xanthine oxidase, NADH/NADPH oxidase, lipoxygenase and nitric oxide synthase, one of the largest factories producing superoxide anion in vivo is the mitochondrion [98]. Via spontaneous or enzymatically catalyzed dismutation, superoxide anion is reduced to hydrogen peroxide. Transition metal, such as iron or copper catalyzed interaction with hydrogen peroxide produces highly toxic hydroxyl radicals. Reactive oxygen species (ROS) have detrimental effect on vascular function through several mechanisms. First, as their direct effect, reactive oxygen species, especially hydroxyl radicals, injure cell membrane and nuclei. Second, by interacting with endogeneous vasoactive mediators formed in endothelial cells, reactive oxygen species modulate vasomotion and atherogenic process. Third, reactive oxygen species peroxide lipid components, leading to formation of oxidized LDL, which is one of the key mediators of atherosclerosis [99]. Whereas native LDL does not cause cholesterol ester accumulation in macrophages, modified LDL by oxidation does [100]. Oxidized-LDL has also been implicated in other mechanisms potentially involved in the development of atherosclerosis, i.e., cytotoxic or chemotactic actions on monocytes and inhibition of macrophage motility [101].

There are a number of clinical markers suggested to assess oxidative stress status (Table 10). Thiobarbituric acid reactive substance (TBARS) is the lipid peroxide derived from unsaturated fatty acids. Isoprotanes such as 8-iso-prostaglandin (PG) F2 alpha or 8-epi-PGF2 alpha are compound formed in vivo by nonenzymatic free radical catalyzed peroxidation of arachidonic acid. TBARS and isoprostanes are both markers for non-specific lipid peroxidation. Elevated urinary 8-iso-PGF2 alpha levels are found in patients with elevated postprandial remnant lipoproteins or in smokers, even in passive smokers. 8-hydroxy-2'-deoxyguanosine (8-OHdG) is a degradation product of DNA oxidation and its measurement provides information on various degrees of oxidative stress in the DNA level.

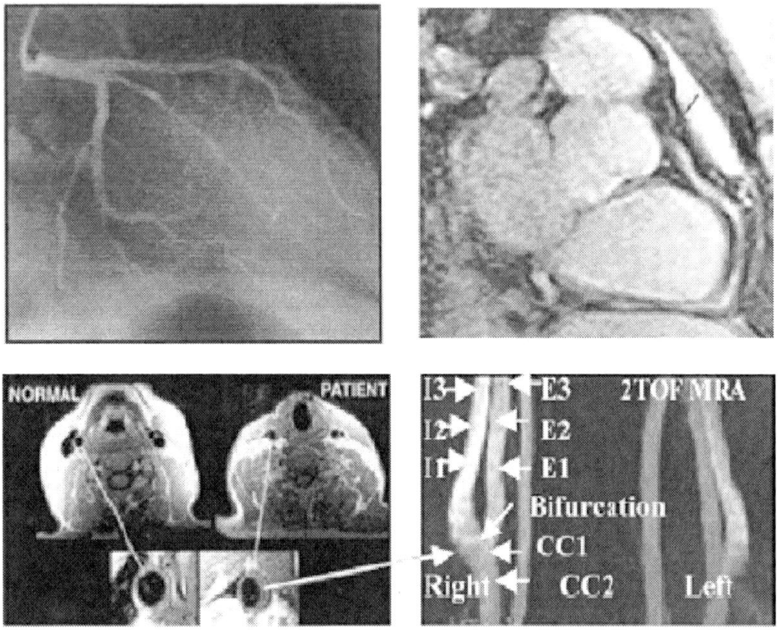

Figure 4. (On top) Coronary angiography showed a stenotic lesion in right coronary artery, and an intravascular magnetic resonance imaging showed a plaque with lipid core in this lesion (right, arrow). (At bottom) A comparison of atherosclerotic carotid artery wall with normal artery (shown in left panel) at different location (shown in right panel). The figure is reproduced with permission from Reference [2].

Table 10. Oxidative stress markers are shown active in atherogenesis

Thiobarbituric acid reactive substance (TBARS)
Oxidized low density lipoprotein(LDL)
Malonic dialdehyde-modified LDL (MDA-LDL)
Oxidized LDL receptor (LOX-1)
8-hydroxy-2-deoxyguanosine (8-OHdG)
8-iso prostaglandin F2α
Anti-oxidized LDL antibody
Oxidized α1-antitrypsin
Tetrahydrobiopterin (BH4)
Asymmetric Di-Methyl arginine(ADMA)

Serum 8-OHdG levels are elevated in smokers and the smoking abstension reduced the levels [102]. Among various lipids peroxidazed, oxidatively modified LDL directly involve in the athelosclerosis process. Thus, the direct measurement of oxidized LDL (Ox-LDL) provides important information regarding oxidative stress in atherosclerosis. On the other hand, specific immunological epitopes expressed on Ox-LDL were found in atherosclerotic lesions both in animals with experimental atherosclerosis and in humans. Expression of such epitopes in vitro can be generated by various procedures, including incubation with endothelial cells and macrophages, oxidation in the presence of copper ions and treatment with malondialdehyde [103]. Ox-LDL can interact with scavenger receptors of monocyte-derived macrophages. It is suggested that these interactions can induce formation of anti-Ox-LDL. Therefore, anti-Ox-LDL can be considered to be a marker of LDL oxidation at the level of tissues or cells. Since anti-Ox-LDL titer in serum of the patients with hypercholesterolemia is not correlated with LDL-cholesterol level but with 8-OHdG, the anti-Ox LDL also seems to be a marker of oxidative DNA damage in dyslipidemic patients [104]. The titer of anti-Ox LDL is associated with short-term lesion progression or regression of atherosclerotic coronary artery disease [105] as well as peripheral artery disease [106, 107], or higher in patients with ACS than in patients with stable angina [108].

9. Guidelines on Lipid Lowering to Physicians, Nurses and Health

The following section is a guide to medical and health care workers who are involved in cardiovascular care and cardioprevention actively. The awareness of both dietary fat intake along with side effects of statins and their safe use is important while practicing them to perform lipid lowering in prospective patient population. The lipid dysfunction includes mainly hypercholesterolemia, hypertriglyceridemia, low HDL-cholesterol and apolipoprotein changes. The target lipid levels (LDL-C <2.5 mmol/L, <3.5 mmol/L, <4.5 mmol/L, and total cholesterol/HDL-C ratio <6.0, <5.0, <4.0) indicate the high risk, moderate risk and low risk of dyslipidemia. The triglycerides >1.7 mmol/L with HDL-C <1.0-1.3 mmol/L indicate metabolic syndrome. Dietary fat intake to improve serum lipid profile includes mainly intake of unsaturated omega-3 and omega-6 fatty acids, antioxidants, L-arginine, folic acid. Of specific mention, ecosapentanoic and docosahexanoic fatty acids, HMG-CoA reductase inhibitors have shown promise in lipid lowering in low risk and moderate risk catagories. Whole foods such as garlic, sprullina, fenugreek, ginko, soy and genestein have been on nonprescription counters as cardioprotective food supplements popular for reducing low and moderate risks. The high risk category (LDL-C <2.5 mmol/L and TC/HDL-C ratio <4.0) needs immediate attention for lipid lowering medication preferably statins. Other choices are BSA, fibrates (benzo-, phenol-), gemfibrozil (400-1200 mg/day) and niacin (1-3 gm). However, health workers need to be aware of the limits of beneficial effects of these supplements and risk of persistent uncontrolled hyperlipidemia. On other hand, statins have gained popularity in acute hyperlipidemia and acute cardiovascular diseases to combat within time but our advice is *KEEP STATINS LOW AS MUCH AS POSSIBLE WITH 'SPRITUAL ACCEPTANCE'*[3] *OF ACTIVE LIFE STYLE*. Moreover, statin therapy is not free from side effects and it needs a right prescription[4] and its safe use under supervision [2]. Present view is ambiguous and frustrating that "low fat diet given to dyslipidemic children appears safe and efficacious when performed under supervision. Medical and nutrition support is necessary to reinforce good dietary behaviors and ensure nutritional adequacy." [2] In 2009 Canadian Cardiovascular Society recommendations [109], new hs-CRP, ApoB/ApoI, TC/HDL-C, carotid intimamedia thickness (CIMT) biomarkers of risk were advocated with heavy emphasis on restricted diet (low sodium and simple sugars, with substitution of unsaturated fats for saturated and trans fats, as well as increased consumption of fruits and vegetables), smoke/alcohol cessation, calorie restriction, least psychological stress based on several recent trials including:

- AFCAPS/TexCAPS;
- West of Scotland Coronary Prevention Study (WOSCOP);
- Anglo-Scandinavian Cardiac Outcomes Trial (ASCOT);
- Heart Protection Study (HPS);
- JUPITER;
- Treating to New Targets (TNT),
- Aggrastat to Zocor (A to Z).

Incremental Decrease in End Points Through Aggressive Lipid Lowering (IDEAL); Assessment of survival and cardiovascular events (AURORA).

[3] http://www.appbrain.com/app/hindu-vrat-calendar/com.pm.HinduVratCalendar.
[4] www.cmaj.ca/cgi/content/full/169/9/921/DC1/

Combinatorial therapy of statin/niacin and acetylsalicylic acid remains inconclusive in following two trials:

Atherothrombosis Intervention in Metabolic Syndrome with low HDL/High Triglyceride and Impact on Global Health Outcomes (AIM-HIGH);

Heart Protection Study 2 Treatment of HDL to Reduce the Incidence of Vascular Events (HPS2-THRIVE).

For full description, readers are referred to read guidelines [109, 110].

Treating beyond Low-Density Lipoprotein Cholesterol

According to NCEP and American Diabetes Association (ADA), LDL-C is the primary therapeutic target in lipid management. As described above, there are several other atherogenic particles which contribute to the CHD risk after LDL-goals are met. Therefore, non-HDL-C is a secondary therapeutic target. [2, 3, 29] The American College of Cardiology (ACC) and ADA statement defines highest risk groups with known cardiovascular disease or patients without cardiovascular disease with DM associated with one or more risk factors such as smoking, hypertension, and family history of premature coronary artery disease(CAD). The 2008 ACC/ADA consensus statement sets specific lipid/lipoprotein goals based on cardiovascular risk based on lipoprotein abnormalities and cardiometabolic risk. The goals for these highest risk patients are LDL-C > 70 mg/dL, non-HDL-C > 100 mg/dL, and apolipoprotein B (Apo B) > 80 mg/dL. [111] Briefly, main points were:

- Each Chylomicrons, VLDL, IDL, LDL and Lp(a) contains a single molecule of apolipoprotein B, making apoB measurement a marker of atherogenic particles. Does not require fasting and assay is standardized
- In several epidemiological and post-hoc analyses of clinical trials apoB has been a better predictor of risk than LDL-C, particularly the on-treatment level of LDL-C
- Once LDL-C is lowered apoB may be a more effective way to assess residual CVD risk and to determine the need for medication adjustments
- The differences between apoB, LDL-C and non-HDL-C are more pronounced in patients with cardiometabolic risk, suggesting apoB is a more useful predictor among these patients

Although statins are the initial drugs of choice, combination therapy may be needed as a strategy to meet lipid goals beyond other factors not just LDL-C target only. Later, in advanced stages, Type I diabetes, nephritic syndrome, polycystic ovarian syndrome happen as major challenges beyond high lipoprotein cholesterol with risk of CVD.

Implications and Futuristic Prospective

In brief, dyslipidemia is a major health hazard since ages and ancient concepts of nutrition appear to be sound for advice on eating natural foods; fruits, vegetables, roots and tubers, sprouted whole grains, nuts, cow milk and curd and honey to combat adverse effects of fried recipes and preserved foods. However, less is known about secrets of high cardiac protection among tribals. There is no advice in modern concepts for fresh animal foods; eggs, fresh fish and meat from running animals commonly consumed by hunter-gatherers, which might be an ignorance in modern literature or a unknown wisdom of 'spiritual health' meaning 'inner senses satisfied with complete acceptance of food available'. However, Rajasic diet advised for fighters is similar to the Kuchirias diet (see Table 3). For example, vegetable GHEE (isolated from butter milk) makes brain strong.

Nowadays we can see success of low-fat dietary life style from a large number of clinical trials. Supervised dietary intervention to reduce low risk of both coronary and carotid artery disease and cardiac prevention are anticipated more rigorously. Large trials have consistently shown a significant benefit of LDL-C intervention but there is a significant residual risk for cardiovascular events especially in high-risk patients with diabetes mallitus. This residual risk is predominantly due to low HDL-C and increased TG in the blood. NCEP has suggested the use of niacin or a fibrate as an additional agent for mixed dyslipidemia.

Ongoing trials will be needed further to demonstrate the incremental cardiovascular disease benefits and safety of combination dietary regimens. In future, statin therapy would be more extensively applied even in normolipidemic patients with new definitions advocated by government or research agencies if patients have other conventional risk factors such as hypertension, diabetes mellitus, or other new factors identified under medical supervision. Furthermore, we may need to use statins to intervene in earlier stage risk factors such as postprandial hyperlipidemia or hyperglycemia, insulin resistant state, masked hypertension, or metabolic syndrome to further reduce mortality or morbidity of coronary artery disease. To identify new risk factors related with hyperlipidemia or cardiovascular risks, is kind of race to pick up the thread early and advise interventions either diet, exercise or life style changes. It remains to establish if prolonged lipid lowering drug treatment is risk free from the any lipid lowering drug-induced toxicity in the body after long duration especially infants, children, adolescents and elderly with different ethnicities. It will be clear if low saturated fat diet in childhood can prevent either CVD in adulthood or obesity due to insulin resistance or enhance endothelial function in teenager boys and girls with clear information how much TC or LDL-C or HDL-C is culprit.

The present chapter suggested the following broader implications: 1. low fat, low sodium with high fiber or PUFA rich diet may cause significant cholesterol and triglyceride lowering in serum but small change, 2. dietary intervention in combination therapy causes lipid lowering and improve serum lipid profile more than dietary intervention alone, 3. dietary low fat energy intake may or may not cause significant change in HDL cholesterol, 4. statin therapy may cause small change in lowering of total and LDL cholesterol in time, and 5. combination of low fat, low salt with omega fatty acids or fibrates may show limited benefit but combined with statin therapy may contribute more to the effect of therapeutic intervention on other factors such as hypertension and mortality from ischemic heart disease. However two issues remain unclear: 1. if prolonged low fat and restricted diet given to low socio-economic population further may add up the risk of malnutrition or immunity loss due to less available energy or resistance to clearance of nutrients as added risk; 2. if low fat diet for long time prescribed may also contribute to any less known adverse effect or limited effect on serum lipids as previous report suggested long back in search of a new lipid regulatory biochemical mechanism(s) [112]. No perfect means of lipid lowering or lipid regulatory mechanism exists today to explain as complete cure without adverse effects. We proposed a novel approach ''keep minimal daily dietary fat intake, low salt, combined omega-3 fatty acids with optimal diuretics and/or betablocker drugs'' may ''keep dyslipidemia in control with normal limits of blood pressure, BMI and serum lipids.'' [113]. Our idea was to highlight the implication of ''low fat optimal drug therapy'' approach to manage dyslipidemia and reduce the mortality from hypertension, coronary or ischemic heart disease and cardiovascular disorders.

The present chapter is focused on two main purposes: first to review abnormal plasma lipids, lipoproteins; different lipid lowering dietary and nutraceutical supplements; possible nutrition therapy or combinatorial drug +

nutrition therapy to observe the subtle benefits or limited cardiac protection by low dietary fat intervention in different social populations at risk; second to establish the ancient concepts on dyslipidemia and explore the less known factors related with long life expectancy with less cardiovascular risk as alternate adjunct means if substitute the prolonged customized intervention (low dietary fat intake, low salt and unwanted maximal statins or betablocker–diuretics therapy) to keep blood pressure and dyslipidemia in normal limits without any side effects or need of identifying new unknown factors. The chapter advocates the importance of new more accurate methods of dyslipidemia monitoring and lipid profile interpretation to avoid drug over dosage. However, lipid profile interpretation has own limitations in different age groups, ethnicity. In our recent study, both man and women were included in a cross-sectional study with no information of their different age groups (adolescent and advanced ages), plasma rennin and ethnicity [112]. Adequate number of multiracial patients was not available whose dietary fat intake was more than 28–30% total intake calories as in Western subjects. This fact restricted our claim that antihypertensive drugs used in our Asian populations may show unlikely any adverse effect(s) on serum lipid profile. Western multiracial population or mixed racial population further warranted to establish relationship between diet modification and drug therapy in dyslipidemic population in terms of serum lipid changes in different ethnicities. Our recent preliminary survey showed clear differences in plasma lipid and lipoprotein levels as shown in Tables 2 and 4 [113].

Other important issue is rapidly changing life styles among inter-country or international migratory high collar employees. Nowadays we observe the success of low-fat dietary life style from a large number of clinical trials but less information on fast migratory populations moving from one corner to other corner of planet. Supervised dietary intervention to reduce low risk of both coronary and carotid artery disease and cardiac prevention are anticipated in such populations or individuals [2, 113]. However, the statins and antihypertensive drugs have gone as established start-up drugs for preventing dyslipidemia and sequential coronary events. Considering their favorable direct effects on blood vessels independently of lipid-lowering effects, statins or betablocker–diuretics nephrotoxicity side effects may prevent their over usage and are now under investigation [2, 114]. However, expectedly the most hopeful endpoints may be primary management and prevention of dyslipidemia with keeping shift of borderline hypertension to normal by intervening statins combined with dietary low salt and low fat intake; and prevention of definitive hypertension or coronary artery disease onset by the use of combined dietary low salt and low fat intake with betablocker–diuretics in patients independently of their serum cholesterol levels. To accomplish these goals, it needs attention for following issues in future [2, 113]. First, we must establish and standardize more accurate body lipid diagnostic methods or in vivo biomarkers for clinical detection of ''dyslipidemia,'' to assess the statin dosage or combinatorial nutrition+ statin option. Similarly, in advanced dyslipidemic population, accurate body lipid and cardiac tissue diagnostic methods are must to establish sequential ''cardiovascular incapability'' before drug intervention to assess the effectiveness of different dosage of drug or need of combinatorial therapies or immediate surgical intervention in time [2]. We believe that degree of association between ''dyslipidemia switch to hypertension or atherosclerosis or diabetes'' with 'abnormal serum lipids and lipoproteins' will serve as a criteria and surrogate endpoint(s) to evaluate the pleiotropic effects of drugs or nutrition in different stages to improve both dyslipidemia and cardiovascular incapability in time. Next, we believe that one must re-evaluate normal and vulnerable limits of serum lipid profiles carefully in population under study with details of life style, social factors, nutrition factors at specific place and environment (independent from specific study report or specific Government guidelines) before intervention of drugs and/or dietary fat restriction in different ethnicities, cultures and age groups with consideration of dyslipidemia resistance in populations, if any, in order to reduce side effects of existing or new antihypertensive drugs. In future, possibly rigorous lipid lowering drug therapy and optimized cardioprotective dietary or nutraceutical supplementation would be more extensively applied even in normolipidemic patients to keep good health with suggestions of dietary low fat intake if they have any conventional risk factors such as hypertension, diabetes mellitus, or others under medical supervision. Furthermore, other alternate cardioprotective medication will be rationalized such as antidiabetic regimens, antihypertensive therapy to intervene in early stage dyslipidemia risks such as postprandial hyperlipidemia or hyperglycemia, insulin resistant state, masked hypertension, or metabolic syndrome to further reduce mortality or morbidity of both coronary artery disease and cardiac heart disease.

More facts of dyslipidemia in infants, children and adulthood will be revealed with clear recognition and definitions of blood lipids with cut-off values in different ethnicities. New foods and dietary regimens have more scope in controlling dyslipidemia to keep away hypertension and cardiovascular disease as indicated in recent study on prospects of NORDIET trial [114].

Conclusions

Present chapter presents a current view on more investigations of dyslipidemia recognition and assessment in different parts of the world and ethnicities, in different age groups. Present state of art on dyslipidemia control and cardiovascular prevention still today remains controversial due to lack of comparisons in different populations and no common agreement on dyslipidemia management. As a result, dyslipidemia risks and consequences remain inconclusive bottlenecks with lot of new side effects of newly introduced pharmaceutical drugs. With growing and advanced techniques of lipid science has opened the new vistas of characterization of different lipid molecule size, visualization, lipoprotein metabolism and regulation, molecular basis of lipoprotein disease during sequential progress of disease either atherogenesis or diabetes with renal complications but medical treatment approaches remain frustrating on reversing disease or complete cure of cardiovascular disease. New dietary and nutrition therapies show some promise because of no side effects due to them used as exclusive nutrition therapy or combinatorial regimens with drugs. In search of better understanding and new diet styles, present chapter surveys the ancient views and life styles with better life expectancy and good health. We put an evidence of Asian Indian tribal practices of traditional diets and their life expectancy without significant cardiac diseases. Awareness of new practices among tribals is reported with no or low risk of dyslipidemia is an exciting and explorable frontline area in both nutrition and social research. New dietary intake information and typical life styles of tribals poses a challenge to both nutritionists and doctors if personal and spiritual acceptance can give some benefits.

References

[1] Fredrickson DS, Lees RS. A system for phenotyping hyperlipoproteinemia. *Circulation* 1965;31:321-327.

[2] Genest J, Frohlich J, Fodor G, McPherson R. Recommendations for the management of dyslipidemia and the prevention of cardiovascular disease: summary of the 2003 update. *CMAJ.* 2003 October 28; 169(9): 921–924.

[3] Kwiterovich PO. Clinical and laboratory assessment of cardiovascular risk in children: Guidelines for screening, evaluation, and treatment. *J. Clin. Lipidol.* 2008 Aug;2(4):248-66.

[4] Sharma R, Katz JK. Sharma R, Katz JK (2011) Evaluation criteria of carotid artery atherosclerosis:Non-invasive multimodal imaging and molecular imaging. *In: Atherosclerosis Disease management. Ed. Suri JS, Kathuria C, Filipo M. Springer Science.XIV, pp 817-878.* http://www.springer.com/biomed/book/978-1-4419-7221-7.

[5] Otvos JD, et al. Measurement issues related to lipoprotein heterogeneity. *Am. J. Cardiol.* 2002;90(8A):22i–29i.

[6] Gordon DJ, et al. High-density lipoprotein cholesterol and cardiovascular disease. Four prospective American studies. *Circulation.*1989;79(1):8–15.

[7] Miller M, et al. Impact of triglyceride levels beyond low-density lipoprotein cholesterol after acute coronary syndrome in the PROVE IT-TIMI 22 trial. *J. Am. Coll. Cardiol.* 2008;51(7):724–730.

[8] Loomis WF. Skin pigment regulation of vitamin D biosynthesis in man. Science 1967; 157:501-506.

[9] Lee RB. What hunters do for a living: A comparative study. In: Lee RB, Devore I, editions, Man the Hunter, Aldine, Chicago 1968;41-43.

[10] Singh RB, De Meester F, Wilczynska A. The Tsim Tsoum approaches for prevention of cardiovascular diseases. Cardiology Research and Practice Volume,2010,ArticleID,824938,18pages,doi:10.4061/2010/824938.

[11] Eaton SB, Eaton SB III, Sinclair AJ, Cordain I, Mann NJ. Dietary intake of long chain polyunsaturated fatty acids during the Paleilithic period. In Simopoulos AP edition. The return of w-3 fatty acids in the food supply. Land based Animal Food Products and their Health Effects. *World Rev. Nutr. Dietetics* 1998; 83: 12-23.

[12] Eaton B. Evolution and cholesterol. *World Rev. Nutr. Diet* 2009; 100: 46-54.

[13] Singh RB, Moshiri M, De Meester F, Juneja L, Muthusamy V, Manoharan S.The evolution of low w-6/w-3 ratio dietary pattern and risk of cardiovascular diseases and diabetes. *J. Altern. Med. Res.* 2011;3(in press).

[14] De Meester F. Wild-type land based foods in health promotion and disease prevention: the LDL-CC:HDL-CC model. In Wild Type Foods in Health Promotion and Disease Prevention, editors Fabien De Meester and RR Watson, Humana Press, NJ 2008, p 3-20.

[15] De Meester F. Progress in lipid nutrition: the Columbus concept addressing chronic diseases. *World Rev. Nutr. Diet* 2009;100: 110-21.

[16] Reddy KK, Papa Rao A and Reddy TPK. Serum Vitamins E, A and lipid peroxidation levels in Kurichias, an Indian tribal population. *Ind. J. Biochem. Biophys.*, 1999;36: 44-50.

[17] Reddy KK, Reddy TPK, Reddy BK, PapaRao A, Singh RB. A sociocultural perspective of centerians in a tribal community of south India. *Help Age India Res. Dev. J.* 2006; 12:30-36.

[18] Singh RB et al. Ancient Nutrition practices in India: A nutrition survey of Kuchirias. *The Open Nutraceutical J.*2011; (in Press).

[19] Third Report of the National Cholesterol Education Program (NCEP) Expert Panel on Detection, Evaluation, and Treatment of High Blood Cholesterol in Adults (Adult Treatment Panel III) final report. *Circulation.* 2002;106(25):3143–3421.

[20] Kaitosaari T, Rönnemaa T, Raitakari O *et al.*, Effect of 7-year infancy-onset dietary intervention on serum lipoproteins and lipoprotein subclasses in healthy children in the prospective, randomized Special Turku Coronary Risk Factor Intervention Project for Children (STRIP) study, *Circulation* 108 (2003), pp. 672–677.

[21] Carson AS. Nutrition therapy for dyslipidemia. Current Diabetes Reports. 2003, 3:397–403.

[22] J.L. Goldstein and M.S. Brown, Molecular medicine: The cholesterol quartet, *Science* 292 (2001), pp. 1310–1312.

[23] D.J. Rader, J. Cohen and H.H. Hobbs, Monogenic hypercholesterolemia: new insights in pathogenesis and treatment, *J. Clin. Invest* 111 (2003), pp. 1795–1803.

[24] M. Arca, G. Zuliani and K. Wilund *et al.*, Autosomal recessive hypercholesterolemia in Sardinia, Italy, and mutations in ARH: a clinical and molecular genetic analysis, *Lancet* 359 (2002), pp. 841–847.

[25] J.D. Horton, J.C. Cohen and H.H. Hobbs, Molecular biology of PCSK9: its role in LDL metabolism, *Trends Biochem. Sci.* 32 (2007), pp. 71–77.

[26] P.O. Kwiterovich, Clinical relevance of the biochemical, metabolic and genetic factors that influence low density lipoprotein heterogeneity, *Am. J. Cardiol.* 90 (suppl 8A) (2002), pp. 30i–48i.

[27] A.D. Sniderman, T. Scantlebury and K. Cianflione, Hypertriglyceridemic hyperapob: the unappreciated atherogenic dyslipoproteinemia in type 2 diabetes mellitus, *Ann. Intern. Med.* 135 (2001), pp. 447–459.

[28] M. Maslowska, H.W. Wang and K. Cianflone, Novel roles for acylation stimulatory protein/C3a desArg: a review of recent in vitro and in vivo evidence, *Vitam Horm* 70 (2005), pp. 309–332.

[29] National Institutes of Health: Third Report of the National Cholesterol Education Program Expert Panel on Detection, Evaluation, and Treatment of High Blood Cholesterol in Adults (Adult Treatment Panel III). Bethesda, MD: National Institutes of Health; 2001. NIH Publication 01-3670. http://www.nhlbi.nih.gov/guidelines/cholesterol/atp3_rpt.htm. Accessed April 1, 2003.

[30] Jenkins DJ, Kendall CW, Faulkner D, *et al.*: A dietary portfolio approach to cholesterol reduction: combined effects of plant sterols, vegetable proteins, and viscous fibers in hypercholesterolemia. *Metabolism* 2002, 51:1596–1604.

[31] Grundy SM: Nutrition and diet in the management of hyperlipidemia and atherosclerosis. In *Modern Nutrition in Health and Disease,* edn 9. Edited by Shils ME, *et al.* Baltimore, MD: Lippincott, Williams and Wilkins; 1999:1199–1216.

[32] Yu-Poth S, Zhao G, Etherton T, *et al.*: Effects of the National Cholesterol Education Program's step I and II dietary intervention programs on cardiovascular disease risk factors: a meta-analysis. *Am. J. Clin. Nutr.* 1999, 69:632–646.

[33] Beauchesne-Rondeau E, Gascon A, Bererson J, Jacques H: Plasma lipids and lipoproteins in hypercholesterolemic men fed a lipid-lowering diet containing lean beef, lean fish or poultry. *Am. J. Clin. Nutr.* 2003, 77:587–593.

[34] Engler MM, Engler MB, Malloy MJ, Paul SM, Kulkarni KR, Mietus-Snyder ML 2005 Effect of docosahexaenoic acid on lipoprotein subclasses in hyperlipidemic children (the EARLY study). *Am. J. Cardiol.* 95:869–871.

[35] Lichenstein AH, Ausman LM, Jalbert Sm, Schaefer EJ: Effects of different forms of dietary hydrogenated fats on serum lipoprotein cholesterol levels. *N. Engl. J. Med.* 1999, 340:1933–1940.

[36] Denke MA, Adams-Huet B, Nguyen AT: Individual cholesterol variation in response to a margarine- or butter-based diet:a study in families. *JAMA* 2000, 284:2740–2747.

[37] Food labeling: trans fatty acids in nutrition labeling, nutrient content claims, and health claims; reopening of the comment period. Food and Drug Administration. *Fed. Reg.* 2002, 67:69171–69172.

[38] Heilbronn LK, Noakes M, Clifton PM: Effect of energy restriction, weight loss and diet composition of plasma lipids and glucose in patients with type 2 diabetes. *Diabetes Care* 1999, 22:889–895.

[39] Ostlund RE: Phytosterols in human nutrition. *Annu Rev. Nutr.* 2002, 22:533–549.

[40] Temme EH, Van Hoydonck PG, Schouten EG, Kesteloot H: Effects of a plant sterol-enriched spread on serum lipids and lipoproteins in mildly hypercholesterolaemic subjects. *Acta Cardiol.* 2002, 57:111–115.

[41] Plat J, van Onselen ENM, van Heugten MMA, Mensink RP: Effects on serum lipids, lipoproteins and fat soluble antioxidant concentrations of consumption frequency of margarines and shortenings enriched with plant stanol esters. *Eur. J. Clin. Nut.* 2000, 54:671–677.

[42] Vanstone CA, Raeini-Sarjaz M, Parsons WE, Jones PJH:Unesterified plant sterols and stanols lower LDL-cholesterol concentrations equivalently in hypercholesterolemic persons. *Am. J. Clin. Nutr.* 2002, 76:1272–1278.

[43] Maki KC, Davidson MH, Umporowicz DM, *et al.*: Lipid responses to plant-sterol-enriched reduced-fat spreads incorporated into a National Cholesterol Education Program Step I diet. *Am. J. Clin. Nutr.* 2001, 74:33–43.

[44] He FJ, Nowson CA, Lucas M, MacGregor GA. Increased consumption of fruit and vegetables is related to a reduced risk of coronary heart disease: meta-analysis of cohort studies. *J. Hum. Hypertens.* 2007;21:717-28.

[45] Heidemann C, Schulze MB, Franco OH et al. Dietary patterns and risk of mortality from cardiovascular disease, cancer, and all causes in a prospective cohort of women. *Circulation.* 2008; 118: 230-7.

[46] Iqbal R, Anand S, Ounpuu S et al. Dietary patterns and the risk of acute myocardial infarction in 52 countries. *Circulation* 2008;118: 1929-37.

[47] Singh RB, DeMeester F, Pella D, Basu TK, Watson RR. Globalization of dietary wild foods protect against cardiovascular disease and all cause mortalities? A scientific statement from International College of Cardiology, Columbus paradigm Institute and International College of Nutrition. *The Open Heart J.* 2009;2:42-45.

[48] Chandalia M, Garg A, Lutjohann D, *et al.*: Beneficial effects of high dietary fiber intake in patients with type 2 diabetes mellitus. *N. Engl. J. Med.* 2000, 342:1392–1398.

[49] Erdman JW: Soy protein and cardiovascular disease: a statement for healthcare professionals from the nutrition committee of the AHA.

American Heart Association Nutrition Committee. *Circulation* 2000, 102:2555–2559.

[50] Food labeling, health claims, soy protein, and coronary heart disease. Food and Drug Administration. *Fed. Reg.* 1999, 57:699–733.

[51] Bhathena SJ, Velasquez MT: Beneficial role of dietary phytoestrogens in obesity and diabetes. *Am. J. Clin. Nutr* 2002, 76:1191–1201.

[52] Teede HJ, Dalais FS, Kotsopoulos D, *et al.*: Dietary soy has both beneficial and potentially adverse cardiovascular effects: a placebo-controlled study in men and postmenopausal women. *J. Clin. Endocrinol. Metab* 2001, 86:3053–3060.

[53] Tonstad S, Smerud K, Hoie L: A comparison of the effects of 2 doses of soy protein or casein on serum lipids, serum lipoproteins, and plasma total homocysteine in hypercholesterolemic subjects. *Am. J. Clin. Nutr.* 2002, 76:78–84.

[54] Hermansen K, Sondergaard M, Hoie L, *et al.*: Beneficial effects of a soy-based dietary supplement on lipid levels and cardiovascular risk markers in type 2 diabetic subjects. *Diabetes Care* 2001, 24:228–233.

[55] Underbakke G: Medical nutrition therapy for cardiovascular disease—beyond the basics. *On the Cutting Edge* 2002, 23:19–22.

[56] Evidence-based nutrition principles and recommendations for the treatment and prevention of diabetes and related complications. American Diabetes Association [no authors listed]. *Diabetes Care* 2003, 26(suppl 1):S51–S61.

[57] Gaede P, Vedel P, Larsen N, *et al.*: Multifactorial intervention and cardiovascular disease in patients with type 2 diabetes. *N. Engl. J. Med.* 2003, 348:383–393.

[58] Sharma R, Singh RB, Moffatt R, Katz J. (2010) Dietary Fat Intake: Promotion of Disease in Carotid Artery Disease: Lipid Lowering Versus Side Effects of Statins. In: Ed. Ronald Ross Watson, Fabien De Meester and Sherma Zibadi Modern Dietary Fat Intakes in Disease Promotion. Nutrition and Health, SpringerLink 2010, Part 3, 151-184.

[59] Bays HE. Safety considerations with omega-3 fatty acid therapy. *Am. J. Cardiol.* 2007;99(6A):35C–43C.

[60] Bays HE, et al. Prescription omega-3 fatty acids and their lipid effects:physiologic mechanisms of action and clinical implications. *Expert Rev. Cardiovasc. Ther.* 2008;6(3):391–409.

[61] Yokoyama M, et al. Effects of eicosapentaenoic acid on major coronary events in hypercholesterolaemic patients (JELIS): a randomized open-label, blinded endpoint analysis. *Lancet.* 2007;369(9567):1090–1098.

[62] Davidson MH, et al. Efficacy and tolerability of adding prescription omega-3 fatty acids 4 g/d to simvastatin 40 mg/d in hypertriglyceridemic patients: an 8-week, randomized, double-blind, placebo-controlled study. *Clin. Ther.* 2007;29(7):1354–1367.

[63] Holman RR, et al. Atorvastatin in Factorial with Omega-3 EE90 Risk Reduction in Diabetes (AFORRD): a randomised controlled trial. *Diabetologia.* 2009;52(1):50–59.

[64] Brown BG. Expert commentary: niacin safety. *Am. J. Cardiol.* 2007;99(6A):32C–34C.

[65] The Coronary Drug Project Research Group: Clofibrate and niacin in coronary artery disease. *JAMA*. 1975;231:360–381.

[66] McKenney J. New perspectives on the use of niacin in the treatment of lipid disorders. *Arch. Intern. Med.* 2004;164(7):697–705.

[67] Ganji SH, et al. Niacin and cholesterol: role in cardiovascular disease(review). *J. Nutr. Biochem.* 2003;14(6):298–305.

[68] Rubic T, et al. Stimulation of CD36 and the key effector of reverse cholesterol transport ATP-binding cassette A1 in monocytoid cells by niacin. *Biochem Pharmacol.* 2004;67(3):411–419.

[69] Ballantyne CM, et al. Comparison of the safety and efficacy of a combination tablet of niacin extended release and simvastatin vs simvastatin monotherapy in patients with increased non-HDL cholesterol (from the SEACOAST I study). *Am. J. Cardiol.* 2008;101(10):1428–1436.

[70] Karas RH, et al. Long-term safety and efficacy of a combination of niacin extended release and simvastatin in patients with dyslipidemia:the OCEANS study. *Am. J. Cardiovasc Drugs*. 2008;8(2):69–81.

[71] Brown BG, et al. Simvastatin and niacin, antioxidant vitamins, or the combination for the prevention of coronary disease. *N. Engl. J. Med.*. 2001;345(22):1583–1592.39.

[72] Taylor AJ, et al. The effect of 24 months of combination statin and extended-release niacin on carotid intima-media thickness: ARBITER 3. *Curr. Med. Res. Opin.* 2006;22(11):2243–2250.

[73] Taylor AJ, et al. HDL-C response to extended release Niacin. *J. Clin. Lipidiol.* 2008;2:285–288.

[74] Karas RH, et al. Long-term safety and efficacy of a combination of niacin extended release and simvastatin in patients with dyslipidemia: the OCEANS study. *Am. J. Cardiovasc. Drugs*. 2008;8(2):69–81.

[75] Brown BG, et al. Simvastatin and niacin, antioxidant vitamins, or the combination for the prevention of coronary disease. *N. Engl. J. Med.*.2001;345(22):1583–1592.39.

[76] Fruchart JC, et al. Peroxisome proliferator-activated receptor-alpha activators regulate genes governing lipoprotein metabolism, vascular inflammation and atherosclerosis. *Curr. Opin. Lipidol.* 1999;10(3): 245–257.

[77] Barbier O, et al. Pleiotropic actions of peroxisome proliferator-activated receptors in lipid metabolism and atherosclerosis. *Arterioscler. Thromb. Vasc. Biol.* 2002;22(5):717–726.

[78] Keech A, et al. Effects of long-term fenofibrate therapy on cardiovascular events in 9795 people with type 2 diabetes mellitus (the FIELD study):randomised controlled trial. *Lancet.* 2005; 366(9500): 1849–1861.

[79] Keech AC, et al. Effect of fenofibrate on the need for laser treatment for diabetic retinopathy (FIELD study): a randomised controlled trial. *Lancet*, 2007;370(9600):1687–1697.

[80] Grundy SM, et al. Clinical management of metabolic syndrome: report of the American Heart Association/National Heart, Lung, and Blood Institute/American Diabetes Association conference on scientific issues related to management. *Circulation.* 2004;109(4):551–556.

[81] Davidson MH. Combination therapy for dyslipidemia: safety and regulatory considerations. *Am. J. Cardiol.* 2002;90(10B):50K–60K.

[82] Grundy SM, et al. Implications of recent clinical trials for the NationalCholesterol Education Program Adult Treatment Panel III guidelines. *Circulation.* 2004;110(2):227–239.

[83] Davidson MH, et al. Safety considerations with fibrate therapy. *Am. J. Cardiol.* 2007;99(6A):3C–18C.

[84] Cannon CP. Combination therapy in the management of mixed dyslipidaemia. *J. Intern. Med.* 2008;263(4):353–365.

[85] Devine PJ, et al. Design and rationale of the ARBITER 6 trial (Arterial Biology for the Investigation of the Treatment Effects of Reducing Cholesterol)-6-HDL and LDL Treatment Strategies in Atherosclerosis (HALTS). *Cardiovasc. Drugs Ther.* 2007;21(3):221–225.

[86] Fung TT, Chiuve SE, McCullough ML,Rexrode KM, Logroscino G, Hu FB. Adherance to DASH- style diet and risk of coronary heart disease and stroke in women. *Arch. Intern. Med.* 2008, 168: 713-720.

[87] Sack FM, Svetkey LP, Vollmer WM, Appel LJ, Bray GA, Harsha D, Obarzanek E, Conlin PR, Miller ER 3rd, Simons-Morton DG, Karajnja

N, Lin PH; DASH-Sodium Collaborative Research Group. Effects on blood pressure of reduced dietary sodium and the Dietary Approaches to Stop Hypertension (DASH) diet. DASH-Sodium Collaborative Research Group. *N. Engl. J. Med.* 2001;344:3-10.

[88] Obarzanek E, Sacks FM, Vollmer WM, Bray GA, Miller ER 3rd , Lin PH, Karanja NM, Most-Windhauser MM, Moore TJ, Swain JF, Bales CW, Proschan MA; DASH Research Group. Effects on blood lipids of a blood pressure-lowering diet: the Dietary Approaches to Stop Hypertension (DASH) Trial. *Am. J. Clin. Nutr* 2001: 74:80-89.

[89] Katcher HI, Legro RS, Kunselman AR et al. The effects of whole grain-enriched hypocaloric diet on cardiovascular disease risk factors in men and women with metabolic syndrome. *Am. J. Clin. Nutr.* 2008,87:79-90.

[90] Singh RB, Beegom R,Verma SP, Haque M, Singh R, Mehta AS, De AK, Kundu S, Roy S, Krishnan A, Simhadri H, Paranjpe NB, Agarwal N. Association of dietary factors and other coronary risk factors with social class in women in five Indian cities. *Asia Pac. J. Clin. Nutr* 2000,9:298-302.

[91] Aratti P, Peluso G, Nicolai R, Calvani M.Polyunsaturated fatty acids:biochemical, nutritional and epigenetic properties. *J. Am. Coll. Nutr* 2004,23:281-302.

[92] Simopoulos AP:"Is Insulin Resistance Influenced by Dietary Linoleic Acid and Trans Fatty Acids?" *Free Radical Biology and Medicine* 1994:17;367-372.

[93] Singh RB, Dubnov G, Niaz MA, Ghosh S, Singh R, Rastogi SS, Manor O, Pella D, Berry EM. Effect of an Indo-Mediterranean diet on progression of coronary disease in high risk patients: a randomized single blind trial. *Lancet* 2002,360:1455-1461.

[94] Trichopoulou A, Bamia C, Trichopoulos D. Anatomy of health effects of Mediterranean diets. Greek epic prospective heart study. BMJ 2009;338: Cite this as: *BMJ* 2009;338:b2337,doi:10.1136/bmj.b2337.

[95] De Lorgeril M, Renaud S, Mamelle N et al. Mediterranean alpha-linolenic acid-rich diet in secondary prevention of coronary heart disease. *The Lancet*, vol. 343, no. 8911, pp. 1454–1459, 1994.

[96] Sharma R. Traditional Therapy Lessons online. http://www.scribd.com/doc/19575964/Traditional-Therapies-Lessons.

[97] Matsuoka H. Endothelial dysfunction associated with oxidative stress in human. *Diabetes Res. Clin.* Pr 2001;54 (supple 2): S65–S72.

[98] Harrison DG. Cellular and molecular mechanism of endothelial cell dysfunction. *J. Clin. Invest.* 1997; 100: 2153–2157.

[99] Steinberg D. Low density lipoprotein oxidation and its pathological significance. *J. Biol. Chem.* 1997; 272:20963–20962.
[100] Fogelman AM, Schechter I, Seager J, et al. Malondialdehyde alteration of low density lipoproteins leads to cholesterol ester accumulation in human monocyte-macrophages. *Proc. Natl. Acad. Sci. USA* 1980; 77:2214–2218.
[101] Quinn MT, Parthasarathy S, Fong LG, et al. Oxidatively modified low density lipoproteins: a potential role in recruitment and retention of monocyte/macrophages during atherosclerosis. *Proc. Natl. Acad. Sci. USA* 1987; 84: 2995–2998.
[102] Inoue T, Hayashi M, Takayanagi K, et al. Oxidative DNA damage is induced by chronic cigarette smoking, but repaired by abstension. *J. Health Sci.* 2003; 29: 217–220.
[103] Palinski W, Ylä-Herttuala S, Rosenfeld ME et al. Antisera and monoclonal antibodies specific for epitopes generated during oxidative modification of low density lipoprotein. *Arteriosclerosis* 1990; 10:325–335.
[104] Inoue T, Inoue K, Maeda H, et al. Immunological response to oxidized LDL occurs in association with oxidative DNA damage independently of serum LDL concentrations in dyslipidemic patients. *Clin. Chim. Acta* 2001; 305: 115–121.
[105] Inoue T, Uchida T, Kamishirado H, et al. Antibody against oxidized low density lipoprotein may predict progression or regression of atherosclerotic coronary artery disease. *J. Am. Coll. Cardiol.* 2001; 37:1871–1876.
[106] Salonen JT, Ylä-Herttuala S, Yamamoto R et al. Autoantibody against oxidized LDL and progression of carotid atherosclerosis. *Lancet* 1992; 339: 883–887.
[107] Bergmark C, Wu R, de Faire RW, et al. Patients with early-onset peripheral vascular disease have increased levels of autoantibodies against oxidized LDL. *Arterioscler. Thromb. Vasc. Biol.* 1995; 15:441–445.
[108] Inoue T, Uchida T, Kamishirado H, et al. Clinical significance of antibody against oxidized low density lipoprotein in patients with atherosclerotic coronary artery disease. *J. Am. Coll. Cardiol.* 2001; 37:775–779.
[109] Genest J, McPherson R, Frohlich J, Anderson T, Campbell N. 2009 Canadian Cardiovascular Society/Canadian guidelines for the diagnosis and treatment of dyslipidemia and prevention of cardiovascular disease

in the adult – 2009 recommendations. *Can. J. Cardiol.* 2009 October; 25(10): 567–579.

[110] Goldstein LB, Bushnell CD, Adams RJ, Appel LJ, Braun LT, Chaturvedi S, Creager MA, Culebras A, Eckel RH, Hart RG, Hinchey JA, Howard VJ, Jauch EC, Levine SR, Meschia JF, Moore WS, Nixon JV, Pearson TA. American Heart Association Stroke Council; Council on Cardiovascular Nursing; Council on Epidemiology and Prevention; Council for High Blood Pressure Research,; Council on Peripheral Vascular Disease, and Interdisciplinary Council on Quality of Care and Outcomes Research. *Stroke.* 2011; 42(2):517-84.

[111] Brunzell JD, Davidson M, Furberg CD et al. Lipoprotein management in patients with cardiometabolic risk: consensus conference report from the American Diabetes Association and the American College of Cardiology Foundation. *Diabetes Care* 2008;31:811-822.

[112] Little P, Girling G, Hasler A, Craven A, Trafford A (1990) The effect of a combination low sodium, low fat, high fibre diet on serum lipids in treated hypertensive patients. *Eur. J. Clin. Nutr.* 44(4):293–300.

[113] Sharma R, Raghuram TC, Rao UB, Moffatt RJ, Krishnaswamy K. The effect of fat intake and antihypertensive drug therapy on serum lipid profile: a cross-sectional survey of serum lipids in male and female hypertensives. *Molec. Cell Biochem*. 343;1-2:37-47.

[114] Adamsson V, Reumark A, Fredriksson IB, Hammarström E, Vessby B, Johansson G, Risérus U. Effects of a healthy Nordic diet on cardiovascular risk factors in hypercholesterolaemic subjects: a randomized controlled trial (NORDIET). *J. Intern. Med.* 2011 Feb; 269(2):150-9.

In: Dyslipidemia
Editors: Z. Ačimovič et.al, pp. 53-85
ISBN: 978-1-62100-917-7

Chapter 2

HDL Heterogeneity and Function: Implications for Therapeutics

Shih Ling Kao[1,2], Ying Jie Chee[2], Sue-Anne Toh[1,2,3]
[1]Division of Endocrinology, University Medicine Cluster,
National University Health System, Singapore
1E Kent Ridge Road, Level 10, Singapore 119228
[2]Yong Loo Lin School of Medicine, National University of Singapore,
1E Kent Ridge Road, Level 10, Singapore 119228
[3]Division of Endocrinology, Department of Medicine,
University of Pennsylvania School of Medicine,
Philadelphia, PA 19104, USA

Abstract

It is recognized that substantial cardiovascular (CV) risk remains even after lowering LDL-C, and that low HDL-C is associated with this residual risk. The anti-atherogenic role of HDL-C is supported by its ability to mediate reverse cholesterol transport, reduce oxidation and inflammation. This has led to substantial interest in HDL-C as a therapeutic target. However, several lines of evidence suggest that HDL-C concentration may be an inadequate representation of its anti-atherogenic effect. Considerable variation is found in cholesterol efflux capacity, a measure of HDL function, amongst individuals with the same HDL-C levels. There have also been situations in which increased CV risk was not observed in individuals with low HDL-C, such as in certain genetic conditions of low HDL-C and ApoA-1. These observations highlight the limitations of HDL-C concentration as a surrogate for anti-atherogenecity. It also points to a need for a more refined measure for use as a therapeutic endpoint. Our ability to quantify the anti-atherogenic effects of HDL-C more accurately will become increasingly important as more HDL-C based therapies are developed.

There are two proposed approaches to address the limitations of measuring HDL-C levels. The first is an assessment of HDL-C heterogeneity. HDL-C exists as a heterogeneous population of subfractions of different densities, charges, sizes and apolipoprotein compostion. These subfractions have been found to correlate differently with CV risk. The second approach is to quantify HDL-C function, using measures such as cholesterol efflux capacity and anti-inflammatory activity. In this chapter, we will review the existing methods of evaluating HDL-C heterogeneity and function. We will also discuss their limitations and potential clinical applications in terms of refinement of cardiovascular risk prediction, and assessment and monitoring of response to HDL-based therapies.

INTRODUCTION

Clinical and epidemiologic studies have consistently shown a strong inverse correlation between HDL-cholesterol (HDL-C) levels and the risk of coronary heart disease [1-3]. Indeed, even in patients treated aggressively with statins to lower levels of the primary target of cholesterol treatment, LDL cholesterol, low HDL-C remains a significant predictor of major cardiovascular events [4]. Hence, it is recognized that substantial cardiovascular disease (CVD) risk remains even after lowering LDL-C, and that low HDL-C is associated with this residual risk. Furthermore, hypercholesterolemic mice with genetic defects in HDL metabolism are markedly atherosclerotic [5], providing compelling evidence that HDL is a key modulator of the disease in animal models. These observations have led to sustained interest in the development of therapies that raise levels of HDL-C.

However, despite the abundant evidence for the inverse association between HDL-C levels and the risk for CVD, certain drugs, such as niacin and fibrates that elevate HDL-C levels, show inconsistent clinical benefit [6, 7]. Moreover, it is not completely understood how HDL in humans interacts with the artery wall to influence the progression or regression of atheroclerosis. The proposed atheroprotective properties of HDL are multifaceted. A central hypothesis revolves around the role of HDL in macrophage reverse cholesterol transport (RCT), the process by which excess cholesterol in foam cells in atheromatous vessels, is effluxed to HDL and ultimately delivered to the liver and excreted into bile [8, 9] (Figure 1). However, whether RCT is the most important mechanism for atheroprotection by HDL remains to be conclusively established. HDL has also been shown to reduce oxidation and inflammation [10-13], inhibit platelet aggregation and coagulation, promote endothelial nitric oxide production [12] and maintain endothelial integrity through anti-apoptotic effects [10, 11] (Figure 2).

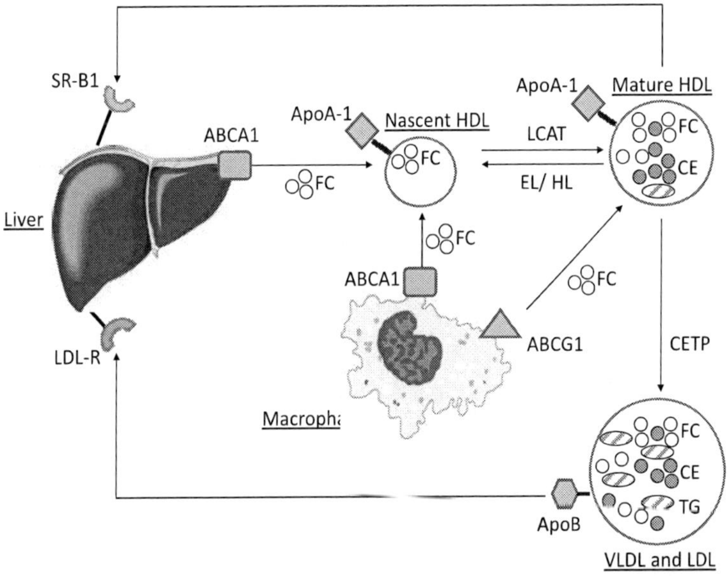

Figure 1. HDL metabolism and reverse cholesterol transport. The liver synthesizes lipid poor ApoA-I, which acquires cholesterol via the hepatocyte ABCA1 transporter. Lipid poor ApoA-I (nascent HDL) also promotes the efflux of free cholesterol from macrophages via ATP-binding casette subfamily A member 1 (ABCA1). Lecithin cholesterol acyltransferase (LCAT) esterifies free cholesterol to cholesteryl esters to form mature HDL, which promotes cholesterol efflux from macrophages via the ATP-binding casette subfamily G member 1 (ABCG1) transporter. Mature HDL can then transfer its cholesterol to the liver directly via cholesteryl ester transfer protein (CETP)-mediated transfer to ApoB-containing lipoproteins, with subsequent uptake by the liver via LDL-receptor. Hepatic cholesterol can then be excreted directly into bile as cholesterol or after conversion to bile acids and, unless reabsorbed by the intestine, is ultimately excreted into feces. HDL can be remodeled by lipases such as hepatic lipase (HL) and endothelial lipase (EL), which hydrolyze triglyceride and phospholipid, respectively.

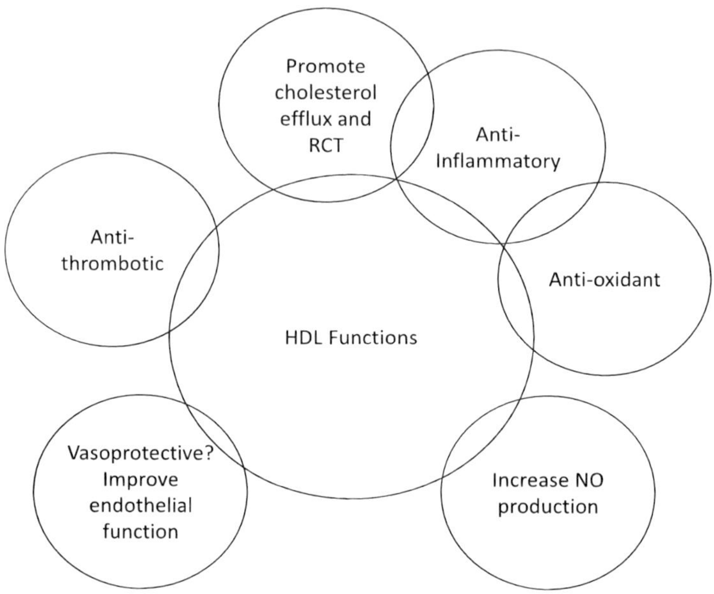

Figure 2. HDL Functions and Properties.Important HDL functional properties include promoting cholesterol efflux and reverse cholesterol transport (RCT), as well as its role in anti-inflammatory processes. These functions are distinct but may be overlapping.

These multiple actions of HDL make it an extremely complex therapeutic target, albeit one with tremendous anti-antiatherogenic potential.

Given this complexity, it is not surprising that several lines of evidence suggest that static measures of plasma steady state HDL-C concentrations alone may not necessarily correlate with HDL function. Plasma HDL-C levels do not provide a measure of RCT rate, nor do they reflect the other properties of HDL; and hence may be an inadequate representation of its potential anti-atherogenic effect. For example, considerable variation is found in cholesterol efflux capacity, amongst individuals with the same HDL-C levels [14, 15]. There have also been situations in which increased CV risk was not observed despite low HDL-C, such as in certain genetic conditions of low HDL-C and ApoA-1 like LCAT deficiency and Apo AI Milano mutation [16, 17]. Torcetrapib, a CETP inhibitor that dramatically increases HDL-C levels was found to be associated with an increase in CV mortality, resulting in the early termination of the ILLUMINATE trial [18]. However, this was attributed to off-target effects of increase in blood pressure and aldosterone caused by torcetrapib, and was not observed with anacetrapib, another CETP inhibitor

[19, 20]. These observations highlight the limitations of depending on HDL-C concentration alone as a surrogate for anti-atherogenecity. It also points to a need for a more refined measure of HDL as a therapeutic endpoint. Our ability to quantify the anti-atherogenic effects of HDL-C more accurately will become increasingly important as more HDL-C based therapies are developed.

There are two proposed approaches to address the limitations of measuring HDL-C levels. The first is an assessment of HDL-C heterogeneity. HDL-C exists as a heterogeneous population of subfractions of different densities, charges, sizes and apolipoprotein compostion. These subfractions have been found to correlate differently with CV risk. The second approach is to quantify HDL-C function, using measures such as cholesterol efflux capacity and anti-inflammatory activity. We will discuss each of these approaches in turn, highlighting potential clinical applications and limitations in terms of refinement of cardiovascular risk prediction, and assessment and monitoring of response to HDL-based therapies.

An Overview of HDL-C Particle Composition and Metabolism

HDL is composed of a phospholipid surface with apolipoproteins and unesterified cholesterol, and a core that carries lipids. Normal HDL also contains high levels of antioxidant molecules, such as paraoxonase/arylesterase1 (PON1) and lecithin cholesterol acyl-transferase (LCAT). Approximately 40-60% of HDL is composed of lipids, such as cholesterol, cholesteryl esters, phospholipids, and triglycerides. The major apolipoproteins of HDL are apoA-I (approximately 70%), apoA-II (20%), apoA-IV, apoCs, apoE, and other apolipoproteins and enzymes (10%) [10]. The other apolipoproteins and nonlipoprotein proteins that associate with HDL are poorly characterized, though recently, Vaisar et al. [21] found forty eight HDL-associated proteins by shotgun proteomics, Interestingly, many of these proteins are not classical apolipoproteins, but rather are complement factors, acute phase proteins or proteases, supporting the concept that HDL's function(s) may not be limited to lipid metabolism.

HDL starts its life cycle as poorly lipidated Apo-A1 (nascent HDL) which collects cellular cholesterol to avoid receptor mediated degradation in the renal tubule [22, 23] (Figure 1). Apo-A1 collects membrane cholesterol through multiple mechanisms, including specific ones requiring physical engagement with transmembrane lipid channels such as ABCA1, ABCG1 and SR-BI [24]. It is believed that ABCA1 transfers phospholipids and cholesterol to the nascent HDL, whereas ABCG1 connects with the larger HDL particles [25]. SR-BI contributes significantly to lipid transfer from cells that are overloaded with cholesterol, such as lipid laden arterial macrophages (ie. the foam cell), and therefore potentially play an important role in atherogenesis [26].

As free (unesterified) cholesterol translocates across the cell membrane, HDL esterifies it by adding a fatty acid chain via the action of LCAT. The non-polar cholesteryl esters are then stored in the particle core. The core expansion leads to compositional maturation, yielding "mature HDL" particles that predominate in plasma HDL. HDL triglyceride content may vary as they undergo lipolysis by hepatic and endothelial lipases to form smaller HDL particles. These sequential changes means that HDL is heterogeneous in size because of variations in the cargo of lipids, apolipoproteins, enzymes, and other functional proteins that can exist on its surface. In plasma, mature HDL switches from cholesterol acquisition to cholesterol delivery. This is achieved mainly via CETP-mediated transfer of cholesterol to triglyceride-rich lipoproteins and via SR-BI mediated unloading of the cholesterol cargo back to the liver. Plasma HDL that eliminates its lipid cargo can initiate a new cycle of peripheral cholesterol acquisition. This process, termed reverse cholesterol transport (RCT), also encompasses the notion that the anti-oxidant and anti-inflammatory effects are linked to the collection and removal of oxidized lipids from cellular membranes and LDL in the arterial wall [27].

Methods for Measurement of HDL Heterogeneity: Potential Clinical Applications and their Limitations

From the above overview of HDL metabolism, it can be appreciated that HDL exists as a heterogeneous mixture of subfractions of different sizes and composition. These subfractions can be separated based on different physiochemical properties such as density, size and charge, with different terminology ascribed to these subfractions depending on the method of separation used.

Analytical Ultracentrifugation

The earliest method of separation of HDL particles was developed by Lindgren et al. in 1951, and involved separation based on density differences [28]. Separation was achieved based on different flotation rates in a high-salt solution. This initially separated HDL into 3 major subclasses: HDL1, HDL2 and HDL3, in order of decreasing size and buoyancy. The HDL2 subclass was subsequently further resolved to yield another 2 subclasses, HDL2a and HDL2b. The optical schlieren profiles obtained after ultracentrifugation can be used to determine mass concentrations of the individual HDL subfractions. This technique formed the basis of the earliest method of HDL quantification before the era of direct homogenous assays [29].

The major subfractions in humans are HDL2 and HDL3. HDL2 is composed of larger cholesterol-rich particles with a density in the range of 1.063-1.125 g/ml, while HDL3 particles are smaller and relatively lipid-poor, with densities in the range of 1.125-1.210 g/ml. HDL1 is a minor subfraction in humans. Several studies have shown that HDL2 and HDL3 have a differential relationship with CHD. However, it remains controversial as to which subfraction is more protective. In a prospective study over 29 years, HDL2 was found to have a stronger inverse relationship with CVD then HDL3 [30]. This echoed findings in several previous studies [31, 32]. However, the reverse was found in other studies such as the Physician's Health Study and the Caerphilly study [33, 34]. The differences in the results could be a result of different assay methods. In view of the conflicting results, routine determination of HDL2 and HDL3 subfractions in clinical practice cannot be presently recommended.

Vertical Analytical Profile

This is another method based on separation by ultracentrifugation, with the difference that the separation is performed in a vertical rotor. This method allows quicker separation of subfractions as it occurs across the horizontal axis of the tube rather then the longer vertical axis. Although the shorter turnaround time makes it more suitable for routine clinical practice then the more laborious analytical ultracentrifugation, there is few data correlating the subfractions obtained from this technique with CVD.

Non-Denaturing Polyacrylamide Gradient Gel Electrophoresis

This technique was developed in 1981 by Blanche et al., and is based on the separation of HDL subfractions by size. HDL is first isolated by ultracentrifugation, and then subjected to gradient gel electrophoresis. This method isolates 5 HDL subfractions of different diameters. They are HDL3c, HDL3b, HDL3a, HDL2a and HDL2b in order of increasing size [35]. There is some, but limited data suggesting that HDL2b is the subfraction that is most strongly and inversely correlated with CVD. Johansson et al. reported that HDL2b was inversely related to the severity and progression of coronary atherosclerosis on angiogram [36].

Two-Dimensional Gel Electrophoresis

This technique separates HDL subfractions by size and charge. HDL typically migrates as an α-band on nondenaturing electrophoresis. Two-dimensional gel electrophoresis further separates HDL into additional subfractions: 1) pre-β-1 HDL, 2) pre-β-2 HDL, 3) pre-α 1,2,3 HDL, and 4) α HDL. These particles have different sizes, chemical composition, and physiological functions. Pre-β-1 HDL is the predominant form of HDL produced by the liver. It is a very small flat disc-like precursor of HDL (diameter 5.6 nm) that contains apo AI, phospholipid, and minimal cholesterol. It is an efficient activator of ATP binding cassette transporter A1 (ABCA1), and is thought to be the most effective HDL particle in terms of promoting cholesterol efflux [37]. Intravenous administration of delipidated HDL particles similar to pre-β-1 HDL in animal models of has been found to lead to regression of atherosclerosis [38].

After incorporation of cholesterol, pre-β-1 HDL enlarges to form α HDL. α HDL can be further resolved by 2D electrophoresis to α-4 HDL, α-3 HDL, α-2 HDL and α-1 HDL, in order of increasing size. α-4 HDL are made up of apo AI, phospholipid and free cholesterol. α-3 HDL and α-2 HDL contain apo AI, apo AII, phospholipid, free cholesterol, cholesterol ester and triglyceride. α-1 HDL, unlike α-2 HDL and α-3 HDL, have very little apo AII. Among the α HDL particles, α-1 HDL and α-2 HDL interact the most efficiently with the liver scavenger receptor B1. [39] The function of Pre-β-2 HDL and pre-α HDL is unclear at present.

The results from several prospective studies suggest that the quantification of HDL subfractions by 2D electrophoresis may be superior compared to HDL measurements alone. In the Framingham Offspring Study, α-1 HDL had the strongest inverse correlation with CVD. CV risk was also inversely correlated with pre-α-3 HDL levels, and positively correlated with pre-β-1 HDL and α-3 HDL levels [40]. Similarly, in the Veterans Affairs HDL Intervention Trial (VA-HIT), α-1 was inversely correlated with CVD, while α-3 was positively correlated. In addition, α-2 was also found to be protective against CVD [41].

Nuclear Magnetic Resonance (NMR) Spectroscopy

The use of NMR spectroscopy to measure lipoprotein particle concentrations was first developed in the 1990s at the North Carolina State University [42]. This method allows the measurements of HDL particle numbers without need for prior physical separation of individual subfractions. It is based upon the principle that different HDL subfractions of varying sizes produce distinct NMR signals at different frequencies, and these signals are directly proportional to the number of particles. The signals used for quantification are emitted from the terminal methyl group protons found in phospholipid, unesterified cholesterol, cholesterol ester and triglyceride, which form the lipoprotein particle. These combine to form a composite signal that is distinct for each subclass of HDL [43]. The overall NMR signal obtained from a plasma sample is a combination of signals from subpopulations of HDL, LDL, VLDL and chylomicrons. This signal is deconvoluted computationally to extract individual subclass amplitudes. Using this method, HDL particles are subclassified as small, medium and large particles, and reported in moles of particles per liter.

There have been several studies published correlating HDL particle number quantified by NMR spectroscopy and CVD. In a case-control study involving the EPIC (European Prospective Investigation into Cancer and Nutrition)-Norfolk cohort, total HDL particle number was found to be an independent predictor of CVD [44]. Total HDL particle number was found to be more strongly correlated with recurrent CHD in the VA-HIT trial, and with carotid intima-media thickness in the MESA (Multi-ethnic Study of Atherosclerosis) trial, compared to HDL cholesterol measurements [45, 46]. In terms of HDL particle size distribution, an inverse relationship between large and small HDL particles have been observed, with large HDL particles being positively associated with CVD and small particles being negatively correlated [47].

Small HDL particles were also found to be an independent predictor of recurrent cardiovascular events in the VA-HIT trial [41]. Unexpectedly, in the JUPITER trial, neither HDL-C nor HDL size, determined by NMR, were associated with residual CVD risk in patients after treated to goal (LDL-C<70 mg/dl) with statins, although statins selectively increase the large HDL2/a-1 subclass [48] The result was in contrast with others' findings [49-51] and the validity of the statistical analysis has been argued [52]. One problem estimating the influence of HDL size on CVD risk is that patients with low concentrations of large and high concentrations of small HDLs also have many other atherogenic factors including increased concentrations of triglycerides, small dense LDL particles and C-reactive protein [53]. Without conducting statistical analyses that address these confounding correlations, misleading conclusions may be reached about the clinical importance and potential functional differences among HDL subclasses [54].

Nonetheless, NMR spectroscopy shows promise for future development for clinical use, as it facilitates accurate and reproducible measurements with a rapid turn-around time. This is especially so as more data correlating these measurements with clinical outcomes are accrued. This method will also contribute to refining HDL measurements in clinical research and provide new information into the physiology of HDL.

Ion Mobility

Ion mobility, also known as gas-phase differential electrical mobility, separates HDL particles based on differences in time of flight through a voltage gradient. The sample is first subjected to an albumin removal step and ultracentrifugation before being introduced into an electrospray unit. The electrosprayed particles then undergo neutralization of charges, separation in an electric field, and are counted by a condensation particle counter [55]. This method currently separates HDL into large HDL particles (HDL 2b) and small HDL particles (HDL 2a and 3).

As this is a relatively new method of HDL separation and quantification, there are limited studies correlating these HDL measurements by this technique and CVD outcomes. The Malmo Diet and Cancer study examined the association of particles measured by ion mobility and CVD risk. It found using principal component analysis that large HDL particles in association with an atherogenic lipoprotein phenotype (increased triglycerides and small LDL particles), was associated with increased CVD risk [56].

Proposed Uniform Nomenclature

The use of different separation methods has led to multiple ways of defining HDL sub-populations, and there is limited comparison of data across methods. A new HDL nomenclature was proposed by Rosenson and colleagues recently [57], with the aim of aligning the HDL subfractions measured using different methods. This new classification is based upon the density and size of HDL and classifies HDL into 5 subgroups: very large HDL (HDL-VL), large HDL (HDL-L), medium HDL (HDL-M), small HDL (HDL-S), and very small HDL (HDL-VS) [57]. The HDL-VL class include HDL-2b species (measured on density gradient ultracentrifugation, gradient gel electrophoresis and ion mobility), α-1 HDL (measured on 2D gel electrophoresis), and large HDL particles (measured on NMR spectroscopy). On the other end of the spectrum, the HDL-VS class include HDL3c (measured on density gradient ultracentrifugation, gradient gel electrophoresis and ion mobility), pre-β-1 HDL (measured on 2D gel electrophoresis) and small HDL particles (measured with NMR spectroscopy). It remains to be seen if this new nomenclature will be adopted in clinical research literature.

Apolipoprotein Based-Classification of HDL Particles: Direct Measures of Apolipoprotein AI and AII

Alaupovic and colleagues proposed classifying lipoproteins based on apolipoprotein composition. Based on this classification, HDL can be subdivided into two groups: LpAI which consists of HDL containing only apo AI, and LpAI:AII which are HDL containing both apo AI and apo AII [58]. These particles can be quantified using immunoprecipitation, two-phase electroimmunoassay, enzyme-linked differential-antibody immunosorbent assay, or differential electroimmunoassay [59]. Differential immunoassay is the most commonly used method in clinical studies. It is still controversial if measuring LpAI and LpAI:AII subfractions are useful clinically. In the Framingham Offspring study and the VA-HIT trial, these measurements were not found to provide prognostic information beyond lipid profile and other CVD risk factors. [40, 41]. It is also unclear which particle, LpAI or LpAI:AII was more inversely correlated with CVD, with some studies suggesting it was LpAI and other studies suggesting both [60, 61]. The classification of HDL particles into subpopulations using this methodology may not sufficiently resolve the heterogeneity of HDL, and this may account for the variable results from clinical studies.

Apo AI is the most abundant protein in HDL, followed by Apo AII. It has been proposed that measuring apolipoprotein levels, especially apo AI, may be superior to measuring HDL-C in terms of CVD risk prediction. This is in view its physiological role in mediating cholesterol efflux via the ABCA1 transporter. Part of the challenge in demonstrating this superiority is that apo AI and HDL-C levels are statistically very highly correlated. There have been a few prospective studies demonstrating that apo AI levels were more predictive, such as in the EPIC-Norfolk cohort and the AMORIS study [62, 63]. It has also been demonstrated that apo AI may undergo oxidation in vivo, which may compromise its anti-inflammatory activity, and ability to mediate reverse cholesterol transport [64, 65]. The measurement of Apo AI is currently not recommended in routine clinical practice. However, there is a potential for it, as well as oxidized apo AI to be developed as prognostic factors for CVD in future. On the other hand, there is still controversy surrounding whether apo aII is pro- or anti-atherogenic. Data from the EPIC-Norfolk cohort and the Ludwigshafen Risk and Cardiovascular Health Study have suggested that it may have a mild anti-atherogenic effect [66, 67]. However, animal studies as well as other human studies have suggested the opposite [68, 69].

Approaches to Assess and Modulate HDL Function

It is evident that HDL particles are heterogeneous in size and composition; and despite the substantial epidemiological data suggesting a cardioprotective role for HDL, much remains unclear about the anti-atheroclerotic and anti-thrombogenic properties of different particles that comprise this class of lipoproteins.

Methods for measurement of HDL subfractions and compositional assays (as described above) may be superior to HDL-C in predicting coronary heart disease risk. We now move on to discuss additional measures for the attributes of HDL, including its role in reverse cholesterol transport, oxidation and inflammation (Figure 2). These aim to provide advantages to current analytical measures of HDL to further understand, validate, and quantify the diverse roles of HDL particles in the atherosclerotic process to improve diagnosis, prevention and treatment of CVDs.

Macrophage Cholesterol Efflux and Reverse Cholesterol Transport

Promotion of cholesterol efflux from macrophages and the return of this excess cholesterol to the liver, bile, and feces, is termed "reverse cholesterol transport" (RCT), and is thought to be one of the most important mechanisms by which HDL protects against atherosclerosis (Figure 1). Measurement of steady state HDL-C level may not be representative of flux through the RCT pathway. Hence, evaluating the flux of cholesterol (the rate and magnitude of inter-compartmental shifts) provides a more dynamic measure of RCT effectiveness and is potentially a more informative way of assessing the efficacy of a novel HDL-targeted intervention.

To assess cellular efflux, donor macrophages are first incubated with 3H-cholesterol [70, 71]. Incubation with a medium containing an "acceptor" (lipid-free apoA-I, isolated HDL, or diluted human serum) is carried out, and after multiple washings, scintigraphy quantifies the radioactivity in the medium and associated with the cells. Cholesterol efflux is then expressed as the amount of label released into the medium divided by the total label present. Acceptors in the medium and donor cells can be manipulated to examine the effects of genetic and pharmacologic manipulation on efflux potential. Manipulation of donor cells may augment cholesterol efflux. For example, mouse peritoneal macrophages over-expressing ABCA1 efflux cholesterol to apoA-I faster than do wild-type control cells, an effect associated with smaller, less complex aortic valvular atherosclerotic lesions in transgenic ABCA1 mice [72]. In humans, investigation of families with ABCA1 mutations demonstrated an inverse correlation between cholesterol efflux and carotid intima-media thickness [73].

The most applicable approach of ex vivo efflux assays to drug development would be to use whole serum or isolated HDL from subjects treated with an experimental drug and test its ability to promote cholesterol efflux from a defined cell system ex vivo. Pre-clinical studies support the concept that augmenting the ability of serum or HDL to promote the efflux of cholesterol from cells may confer an atheroprotective benefit. Administration of recombinant apoA-I Milano or oral D-4F, a small apoA-I mimetic peptide, significantly increased the cholesterol efflux-promoting capacity of plasma taken from treated apoE-deficient mice [74]. These ex vivo findings were associated with decreased macrophage and lipid content of aortic atheroma, indicating enhanced mobilization of tissue cholesterol. Interestingly, treatment with the CETP inhibitor, torcetrapib has been shown to enhance efflux capacity of HDL, whether normalized to volume of serum or to HDL mass [75].

However, until recently, there was scarcity of data that link serum efflux capacity to atherosclerosis or cardiovascular events in humans. A significant step forward in this regard was made when Rader and colleagues measured the ability of human serum HDL to promote cholesterol efflux from cultured macrophage foam cells in two large, independent groups of subjects, and correlated this with coronary disease status [14]. It is important to note that the association was inverse, which is consistent with studies in animals that show that HDL is atheroprotective because it promotes cholesterol efflux from macrophages. Further, the ability of serum HDL to promote cholesterol efflux was not determined simply by measuring HDL-C levels, as the association between efflux capacity and the risk of coronary artery disease remained significant even after adjustments for levels of HDL-C and apo A-I. These observations provide important evidence that the ability to promote cholesterol efflux from macrophage foam cells is a key property that explains, at least in part, the inverse relationship between HDL and the risk of atherosclerotic coronary artery disease in humans. This supports the concept that HDL efflux capacity is a measure of HDL function that is relevant to the pathogenesis of atherosclerosis and that dysfunctional HDL contributes to the risk of coronary disease. However, a limitation of efflux potential as a surrogate measure is its inability to assess terminal segments of the RCT pathway or effects of interventions that are directed toward the cells themselves. There is a need to determine the different pathways that contribute to total efflux to serum and how these vary among individuals. Overall, there is also a need for additional data linking serum efflux capacity to atherosclerosis progression and clinical outcomes.

Further, robust and sensitive methods for assessing integrated RCT in humans are needed in order to advance insights gained in animals into the human realm and to assess novel therapies targeted toward HDL and RCT. One potential approach is simply to measure the mass of fecal sterol excretion as a surrogate for RCT. For example, an acute intravenous bolus infusion of pro-apoA-I in humans was found to result in a significant increase in fecal sterol excretion, suggesting promotion of RCT [76]. However, this approach is not macrophage specific, is unlikely to be very sensitive, and may have limited utility in the chronic steady-state setting due to counter-regulatory pathways involved in biliary cholesterol excretion and fecal sterol absorption.

A steady-state isotope dilution technique has been developed that involves the intravenous infusion of cholesterol labeled with stable-isotope, which is diluted by the efflux of endogenous (unlabeled) cholesterol from tissues into plasma. Presentation of preliminary human studies using 13C-cholesterol demonstrated feasibility, revealing that an 18-h infusion period was required to achieve plateau tracer levels and reproducible measurements of tissue cholesterol efflux [77]. No primary data are yet available in the peer-reviewed literature, but a more detailed discussion of this method can be found in a review [13]. It remains to be established whether this method will have utility in assessing RCT in humans following therapeutic intervention. However, its ability to differentiate between hepatic and peripheral (non-hepatic) tissues as the source of cholesterol efflux has not been definitively proven. Furthermore, as with fecal sterol excretion, this method is not macrophage specific. Nevertheless, it represents an interesting new approach to this problem.

An alternative approach would be to load peripheral tissues with a cholesterol-like molecule that is not endogenously synthesized but is effluxed and transported similarly as cholesterol and then to follow the rate of disappearance of this tracer from tissues that can be sampled repeatedly (circulating cells, skin, adipose) as well as the rate of excretion in the feces. As with the above methods, however, this is not macrophage specific. Ideally, one would like to label cholesterol specifically in arterial wall macrophages in humans and trace its efflux to plasma through to fecal excretion, a goal that is probably unattainable. However, there may be approaches to selectively labeling cholesterol in macrophages using an in vivo targeting approach that could be adapted to RCT studies in humans. Development of a robust assay for RCT in humans, ideally from the macrophage, will be critical to the development of novel therapeutics that may increase RCT as an antiatherogenic strategy.

Indexes of Anti-Inflammatory and Antioxidant Activity

Normal HDL is an effective antioxidant and anti-inflammatory molecule, by virtue of the proteins and lipids associated with HDL that contribute to its anti-inflammatory capacity. HDL particles can exert potentially antiatherogenic effects independent of cholesterol efflux and centripetal transport, including inhibiting lipid oxidation, impairing leukocyte adhesion and monocyte activation, and preventing endothelial cell damage and death [12, 78]. These anti-inflammatory activities of HDL have generated significant excitement, prompted by evidence implicating LDL modification and monocyte migration, which are critical processes for atherogenesis.

Monocyte Chemotactic Activity

To characterize the anti-inflammatory or proinflammatory potential of HDL, a cell-based assay measures the effect of HDL on monocyte chemotactic activity, which is the propensity of monocytes to migrate into the subendothelial space [79]. This co-culture model provides a simulated arterial wall consisting of a confluent human aortic endothelial cell monolayer and smooth muscle cell multilayer separated by a layer of collagen. The endothelial cell monolayer is first incubated with standard control LDL or oxidized phospholipids, stimulants of monocyte chemotaxis, in the presence or absence of test HDL.

Human peripheral blood monocytes are subsequently added to the endothelial side of the coculture and visualized by the addition of antimonocyte antibodies. Finally, monocytes that migrated into the subendothelial space are counted.

Results obtained from coculture incubation with test HDL can be divided by those following addition of control LDL alone to yield an "HDL inflammatory index," with values above 1 indicating proinflammatory HDL and values below 1 denoting anti-inflammatory HDL [80].

An interesting human study [80] compared HDL inflammatory indexes in healthy control subjects and patients with coronary artery disease (CAD) before and after statin therapy. High-density lipoprotein cholesterol taken from age- and gender-matched healthy control patients decreased monocyte chemotaxis, yielding a mean inflammatory index of 0.38. Patients with CAD, however, exhibited proinflammatory HDL augmenting monocyte transmigration more than control LDL alone, with a mean index of 1.38. Administration of simvastatin significantly reduced the inflammatory index to 1.08, suggesting improvement of HDL anti-inflammatory function with statin therapy.

In addition, the study evaluated monocyte chemotaxis in CAD patients with elevated HDL-C levels (mean 95 mg/dl) in an attempt to explain the apparent discrepancy. The HDL obtained from this group was proinflammatory with an index of 1.28, highlighting the limitations of HDL-C alone and the importance of functional metrics. Although this assay is conceptually attractive and has been validated in several settings, it may be cumbersome to use as a routine assay for assessing HDL anti-inflammatory function.

Anti-Oxidant Activity

Another aspect of HDL function relevant for atherogenesis is the fact that lipid oxidation products in HDL inhibit antioxidant and anti-inflammatory enzymes associated with HDL, including PON1 and glutathione selenoperoxidase [81]. One recently developed cell-free assay [82] measures integrated HDL antioxidant potential, using physiologic phospholipids known to participate in LDL modification. In the presence of dichlorofluorescein, a fluorescent marker of lipid oxidation products, HDL is added to various phospholipids found in oxidized LDL to determine its ability to inactivate or prevent the formation of biologically active oxidized phospholipids. After the reagents are combined, spectroscopy permits quantification of net oxidation, with diminished fluorescence intensity signaling fewer oxidized phospholipids and suggesting a more antiatherogenic HDL. High-density lipoprotein from CAD patients yields significantly higher levels of oxidation products compared with that of healthy control patients [82]. Treatment with simvastatin patients but fails to restore its antioxidant ability [80]. In another small study of immigrants from South Asia living in the USA, carotid intima-media thickness substantially correlated with dysfunctional HDL after adjusting for age, family history of CVD, and hypertension [83]. This assay is an interesting candidate for validation and application to high-throughput assessment of HDL antioxidant function in response to therapeutic intervention.

Expression of Endothelial Expression Molecules

Studies of endothelial expression of adhesion molecules may provide additional methods for evaluating HDL-directed therapeutics. To gain entry into the subendothelial space, monocytes must first arrest their passage through the circulation. Adhesion molecules presented on the luminal surface of endothelial cells slow neighboring monocytes, first via loose tethering mediated by E-selectin and then by tight association of endothelial intercellular adhesion molecule (ICAM)-1 and vascular cell adhesion molecule (VCAM)-1 with leukocyte integrins [84, 85]. Expression of adhesion molecules represents a dynamic process, activated by inflammatory cytokines [86] and vascular injury [87] and up-regulated in atherosclerosis in both animals [88] and humans [89, 90.]. As data implicate increased endothelial expression of adhesion molecules in atherosclerotic disease [91], in vitro assays have been developed to evaluate the effect of HDL on this particular facet of endothelial activation (92). The original experiment using human umbilical endothelial cells (HUVECs) demonstrated the ability of HDL obtained from healthy donors to inhibit cytokine-mediated protein expression of VCAM-1, ICAM-1, and E-selectin [92]. Subsequent in vitro studies have established the ability of native and reconstituted HDL to down-regulate inducible expression of adhesion molecules [93, 94]. Importantly, suppression of endothelial expression of adhesion molecules in vivo, which has been associated with decreased atherosclerosis [95], appears to correlate with in vitro findings. However, the reliability and reproducibility of the in vitro assay of inhibition of endothelial adhesion molecule expression as a surrogate measure of the antiatherogenic functionality of HDL remains to be established, and standardized and validated assays are needed.

NO Production and Endothelial Function

Quiescent endothelium produces endothelial nitric oxide (NO), which acts to inhibit cellular pathways of inflammation, proliferation, and thrombosis. HDL-C has been shown to promote endothelial generation of NO in vitro and improve endothelial function and arterial vasoreactivity in vivo [12]. Thus, assaying endothelial NO production in response to HDL could provide the basis of an in vitro proxy of endothelial function. In principle, basal endothelial production of nitrite and nitrate is generated by incubating endothelial cells with L-arginine, the substrate of nitric oxide synthase (NOS), and stimulated production is achieved through simultaneous addition of an NOS agonist. The NO synthase activity may be determined by measuring NOS conversion of [3H]Larginine to [3H]L-citrulline via scintillation counting [96] or by quantifying production of nitrite and nitrate via ozone chemiluminescence [97].

This approach has been used to assess the efficacy of HDL-based interventions. For example, the apoA-I mimetic L-4F increased endothelial cell NO generation under basal and stimulated conditions [98]. A subsequent study [99] correlated these in vitro findings with an improved vasodilatory response to acetylcholine in hypercholesterolemic, LDL receptor-null mice. Evaluating upstream effects of isolated HDL after an intervention on NOS activation and localization in caveolae, specialized plasma membrane microdomains containing a variety of signal transduction molecules, may also prove informative [100]. Finally, apart from preserving NO activity, HDL exerts other putatively atheroprotective effects on endothelial cells, promoting their proliferation and migration [101] and inhibiting their apoptosis [102]. Widespread application of these in vitro assays as a surrogate measure in HDL-targeted therapy development, will require further validation in intervention studies to ensure its in vivo clinical relevance following pharmacotherapy.

Vasoreactivity is the most common method of assessing endothelial function in vivo and is usually studied in humans using ultrasound imaging of the brachial artery at baseline and during reactive hyperemia [103]. Longitudinal images are initially obtained after a 10-min equilibration period with patients in a fully recumbent position. A blood pressure cuff is then placed proximal to the transducer on the upper arm and inflated to above systolic pressure. After 5 min of occlusion, the blood-pressure cuff is deflated and the brachial artery scanned continuously for 60 s. Finally, flow-mediated dilation (FMD) is calculated as the maximum percent increase in brachial artery diameter after 1 min of reactive hyperemia. In observational studies of patients with established CAD [104, 105], patients with atherosclerotic risk factors [106, 107], and healthy subjects [108], the extent of FMD has consistently correlated with HDL-C levels. Studies of HDL-directed therapies have utilized FMD as a proxy of endothelial protection [109]. Although FMD has been suggested to have prognostic value [110], change in FMD following pharmacotherapy has never been correlated with cardiovascular outcomes. Ongoing studies will hopefully further elucidate and validate techniques to assess endothelial effects mediated by HDL to clarify their utility as reliable biomarkers in the setting of pharmacotherapy.

Antiplatelet and Antithrombotic Activity

HDL-C may provide additional cardiovascular benefits by antagonizing platelet activity and inactivating the coagulation cascade, both through direct inhibition and via restoration of endothelial function [12]. Employing a rat model of aortic thrombosis, one pre-clinical trial [111] demonstrated that administration of apoA-I Milano significantly delayed time to thrombus formation, inhibited platelet aggregation, and reduced the weight of thrombus produced. In a human study of hyperlipidemic and normolipidemic patients without significant coronary risk factors [112], high HDL-C levels were associated with reduced thrombus in an ex vivo model. Clot formation, evaluated by exposing porcine aortic intima to flowing, nonanticoagulated venous blood from test subjects, varied inversely to HDL-C in univariate and multivariable analyses.

Within the intricate cascade of hemostasis and thrombosis, a multiplicity of interacting factors complicates determination and prioritization of clinically relevant, HDL specific pathways. Studies thus far [113, 114] implicate the serine protease protein C, a major physiologic anticoagulant that inactivates factors Va and VIIIa in plasma. Using a modified prothrombin time clotting assay, one experiment [115] demonstrated that HDL significantly enhanced inactivation of coagulation factor Va by activated protein C and protein S, and levels of apoA-I correlated with the anticoagulant response. There is also evidence to suggest that HDL may exert antithrombotic activity via inhibition of tissue factor expression, factor X activation, and plasminogen activator inhibitor secretion [116-118]. Overall, surrogate measures of HDL antiplatelet and antithrombotic functionality remain underdeveloped, largely due to the absence of consensus regarding the clinically relevant atheroprotective pathways, and none have been consistently employed or correlated with plaque burden or morphology. However, it is important to note the accumulating evidence implicating platelet activation and thrombin generation in the chronic progression of atheroma [119-121], as well as the final, catastrophic event of plaque rupture leading to acute coronary syndrome and other atherothrombotic diseases. These data support a critical role of platelets and the coagulation cascade in the ongoing progression of atherosclerosis and highlight the importance of elucidating HDL-mediated pathways and developing reliable metrics of antiplatelet and antithrombotic activity.

CONCLUSION

We have reviewed the approaches currently available for evaluating HDL heterogeneity and function. The dichotomy between quantity and quality is a somewhat loose one; mainly to help conceptualize the strategies targeting HDL for therapeutic purposes. At present, which of these strategies will prove to be most effective remains unknown and is an intense area of ongoing investigation. Alongside with the development of therapeutics must be the development of assays for evaluating HDL function that can be adapted for widespread clinical use. Clearly, the current practice of measuring a static HDL-C level is not sufficiently comprehensive. It would be helpful to have a validated, widely available assay of HDL function that can be used in clinical trials to assess new compounds as they enter into clinical development, as not all HDL therapeutics increase HDL-C levels and even those that do may not necessarily improve HDL function. Robust assays that can evaluate the function of HDL, thus supplementing the measurement of HDL-C levels in the clinic, will hopefully leave us with multiple, possibly synergistic, HDL therapeutics that will help reduce residual cardiovascular risk in the population.

REFERENCES

[1] Cooney MT, Dudina A, De Bacquer D, Wilhelmsen L, Sans S, Menotti A, De Backer G, Jousilahti P, Keil U, Thomsen T, Whincup P, Graham IM; SCORE investigators. HDL cholesterol protects against cardiovascular disease in both genders, at all ages and at all levels of risk. *Atherosclerosis*. 2008. 206:611-616.

[2] Gordon T, Castelli WP, Hjortland MC, Kannel WB, Dawber TR. High density lipoprotein as a protective factor against coronary heart disease. The Framingham Study. *Am. J. Med.* 1977. 62:707-714.

[3] Emerging Risk Factors Collaboration, Di Angelantonio E, Sarwar N, Perry P, Kaptoge S, Ray KK, Thompson A, Wood AM, Lewington S, Sattar N, Packard CJ, Collins R, Thompson SG, Danesh J. Major lipids, apolipoproteins, and risk of vascular disease. *JAMA*. 2009. 302:1993-2000.

[4] Barter P, Gotto AM, LaRosa JC, Maroni J, Szarek M, Grundy SM, Kastelein JJ, Bittner V, Fruchart JC; Treating to New Targets Investigators. HDL cholesterol, very low levels of LDL cholesterol, and cardiovascular events. *N. Engl. J. Med.* 2007. 357:1301-1310.

[5] Tall AR, Yvan-Charvet L, Terasaka N, Pagler T, Wang N. HDL, ABC transporters, and cholesterol efflux: implications for the treatment of atherosclerosis. *Cell Metab.* 2008. 7:365-375.

[6] Joy T, Hegele RA. Is raising HDL a futile strategy for atheroprotection? *Nat. Rev. Drug Discov*. 2008. 7:143-155.

[7] NHLBI, 2011. NIH stops clinical trial on combination cholesterol treatment. [press release], 26 May 2011, Available at: <http://www.nih.gov/news/health/may2011/nhlbi-26.htm> [Accessed 28 August 2011]

[8] Lewis GF. Determinants of plasma HDL concentrations and reverse cholesterol transport. *Curr. Opin. Cardiol.* 2006. 21:345-352.

[9] Mahley RW, Huang Y, Weisgraber KH. Putting cholesterol in its place: apoE and reverse cholesterol transport. *J. Clin. Invest.* 2006. 116:1226-1229.

[10] Navab M, Reddy ST, Van Lenten BJ, Fogelman AM. HDL and cardiovascular disease: atherogenic and atheroprotective mechanisms. *Nat. Rev. Cardiol.* 2011. 8:222-232.

[11] Assmann G, Gotto AM Jr. HDL cholesterol and protective factors in atherosclerosis. *Circulation.* 2004. 109:III8-III14.

[12] Mineo C, Deguchi H, Griffin JH, Shaul PW. Endothelial and antithrombotic actions of HDL. *Circ Res.* 2006. 98:1352-1364.

[13] deGoma EM, deGoma RL, Rader DJ. Beyond high-density lipoprotein cholesterol levels evaluating high-density lipoprotein function as influenced by novel therapeutic approaches. *J. Am. Coll Cardiol.* 2008. 51:2199-2211.

[14] Khera AV, Cuchel M, de la Llera-Moya M, Rodrigues A, Burke MF, Jafri K, French BC, Phillips JA, Mucksavage ML, Wilensky RL, Mohler ER, Rothblat GH, Rader DJ. Cholesterol efflux capacity, high-density lipoprotein function, and atherosclerosis. *N. Engl. J. Med.*. 2011. 364(2):127-135.

[15] de la Llera-Moya M, Drazul-Schrader D, Asztalos BF, Cuchel M, Rader DJ, Rothblat GH. The ability to promote efflux via ABCA1 determines the capacity of serum specimens with similar high-density lipoprotein cholesterol to remove cholesterol from macrophages. *Arterioscler Thromb. Vasc. Biol.* 2010. 30:796-801.

[16] Kuivenhoven JA, Pritchard H, Hill J, Frohlich J, Assmann G, Kastelein J. The molecular pathology of lecithin:cholesterol acyltransferase (LCAT) deficiency syndromes. *J. Lipid Res.* 1997. 38:191-205.

[17] Franceschini G, Sirtori CR, Capurso A 2nd, Weisgraber KH, Mahley RW. A-IMilano apoprotein. Decreased high density lipoprotein cholesterol levels with significant lipoprotein modifications and without clinical atherosclerosis in an Italian family. *J. Clin. Invest.* 1980. 66:892-900.

[18] Barter PJ, Caulfield M, Eriksson M, Grundy SM, Kastelein JJ, Komajda M, Lopez-Sendon J, Mosca L, Tardif JC, Waters DD, Shear CL, Revkin JH, Buhr KA, Fisher MR, Tall AR, Brewer B; ILLUMINATE Investigators. Effects of torcetrapib in patients at high risk for coronary events. *N. Engl. J. Med.* 2007. 357:2109-2122.

[19] Rader DJ. Illuminating HDL--is it still a viable therapeutic target? *N. Engl. J. Med.* 2007. 357:2180-2183.

[20] Cannon CP, Shah S, Dansky HM, Davidson M, Brinton EA, Gotto AM, Stepanavage M, Liu SX, Gibbons P, Ashraf TB, Zafarino J, Mitchel Y, Barter P; Determining the Efficacy and Tolerability Investigators. Safety of anacetrapib in patients with or at high risk for coronary heart disease. *N Engl. J. Med.* 2010. 363:2406-2415.

[21] Vaisar T, Pennathur S, Green PS, Gharib SA, Hoofnagle AN, Cheung MC, Byun J, Vuletic S, Kassim S, Singh P, Chea H, Knopp RH, Brunzell J, Geary R, Chait A, Zhao XQ, Elkon K, Marcovina S, Ridker P, Oram JF, Heinecke JW. Shotgun proteomics implicates protease inhibition and complement activation in the antiinflammatory properties of HDL. *J. Clin. Invest.* 2007. 117:746-756.

[22] Hammad SM, Barth JL, Knaak C, Argraves WS. Megalin acts in concert with cubilin to mediate endocytosis of high density lipoproteins. *J Biol Chem.* 2000. 275:12003-12008.

[23] Graversen JH, Castro G, Kandoussi A, Nielsen H, Christensen EI, Norden A, Moestrup SK. A pivotal role of the human kidney in catabolism of HDL protein components apolipoprotein A-I and A-IV but not of A-II. *Lipids*. 2008. 43:467-470.

[24] Zannis VI, Chroni A, Krieger M. Role of apoA-I, ABCA1, LCAT, and SR-BI in the biogenesis of HDL. *J. Mol. Med. (Berl).* 2006. 84:276-294.

[25] Vaughan AM, Oram JF. ABCA1 and ABCG1 or ABCG4 act sequentially to remove cellular cholesterol and generate cholesterol-rich HDL. *J. Lipid Res.* 2006. 47:2433-2443.

[26] Yancey PG, Kawashiri MA, Moore R, Glick JM, Williams DL, Connelly MA, Rader DJ, Rothblat GH. In vivo modulation of HDL phospholipid has opposing effects on SR-BI- and ABCA1-mediated cholesterol efflux. *J. Lipid Res.* 2004. 45:337-346.

[27] Navab M, Ananthramaiah GM, Reddy ST, Van Lenten BJ, Ansell BJ, Fonarow GC, Vahabzadeh K, Hama S, Hough G, Kamranpour N, Berliner JA, Lusis AJ, Fogelman AM. The oxidation hypothesis of atherogenesis: the role of oxidized phospholipids and HDL. *J. Lipid Res.* 2004. 45:993-1007.

[28] Lindgren FT, Elliott HA, Gofman JW. The ultracentrifugal characterization and isolation of human blood lipids and lipoproteins, with applications to the study of atherosclerosis. *J. Phys. Colloid Chem.* 1951. 55:80-93.

[29] Anderson DW, Nichols AV, Pan SS, Lindgren FT. High density lipoprotein distribution. Resolution and determination of three major components in a normal population sample. *Atherosclerosis.* 1978. 29:161-179.

[30] Williams PT, Feldman DE. Prospective study of coronary heart disease vs. HDL2, HDL3, and other lipoproteins in Gofman's Livermore Cohort. *Atherosclerosis.* 2011. 214:196-202.

[31] Salonen JT, Salonen R, Seppänen K, Rauramaa R, Tuomilehto J. HDL, HDL2, and HDL3 subfractions, and the risk of acute myocardial infarction. A prospective population study in eastern Finnish men. *Circulation.* 1991. 84:129-139.

[32] Gofman JW, Young W, Tandy R: Ischemic heart disease, atherosclerosis, and longevity. *Circulation.* 1966. 34:679-697.

[33] Stampfer MJ, Sacks FM, Salvini S, Willett WC, Hennekens CH. A prospective study of cholesterol, apolipoproteins, and the risk of myocardial infarction. *N. Engl. J. Med.* 1991. 325:373-381.

[34] Yu S, Yarnell JW, Sweetnam P, Bolton CH. High density lipoprotein subfractions and the risk of coronary heart disease: 9-years follow-up in the Caerphilly Study. *Atherosclerosis.* 2003. 166:331-338.

[35] Blanche PJ, Gong EL, Forte TM, Nichols AV. Characterization of human high-density lipoproteins by gradient gel electrophoresis. *Biochim. Biophys. Acta.* 1981. 665:408-419.

[36] Johansson J, Carlson LA, Landou C, Hamsten A. High density lipoproteins and coronary atherosclerosis. A strong inverse relation with the largest particles is confined to normotriglyceridemic patients. *Arterioscler Thromb.* 1991. 11:174-182.

[37] Asztalos BF, de la Llera-Moya M, Dallal GE, Horvath KV, Schaefer EJ, Rothblat GH. Differential effects of HDL subpopulations on cellular ABCA1- and SR-BI-mediated cholesterol efflux. *J. Lipid Res.* 2005. 46:2246-2253.

[38] Sacks FM, Rudel LL, Conner A, Akeefe H, Kostner G, Baki T, Rothblat G, de la Llera-Moya M, Asztalos B, Perlman T, Zheng C, Alaupovic P, Maltais JA, Brewer HB. Selective delipidation of plasma HDL enhances reverse cholesterol transport in vivo. *J. Lipid Res.* 2009. 50:894-907.

[39] Asztalos BF, de la Llera-Moya M, Dallal GE, Horvath KV, Schaefer EJ, Rothblat GH. Differential effects of HDL subpopulations on cellular ABCA1- and SR-BI-mediated cholesterol efflux. *J. Lipid Res.* 2005. 46:2246-2253.

[40] Asztalos BF, Cupples LA, Demissie S, Horvath KV, Cox CE, Batista MC, Schaefer EJ. High-density lipoprotein subpopulation profile and coronary heart disease prevalence in male participants of the Framingham Offspring Study. *Arterioscler Thromb. Vasc. Biol.* 2004. 24:2181-2187.

[41] Asztalos BF, Collins D, Cupples LA, Demissie S, Horvath KV, Bloomfield HE, Robins SJ, Schaefer EJ.

[42] Value of high-density lipoprotein (HDL) subpopulations in predicting recurrent cardiovascular events in the Veterans Affairs HDL Intervention Trial. *Arterioscler Thromb. Vasc. Biol.* 2005. 25:2185-2191.

[43] Otvos JD, Jeyarajah EJ, Bennett DW, Krauss RM. Development of a proton nuclear magnetic resonance spectroscopic method for determining plasma lipoprotein concentrations and subspecies distributions from a single, rapid measurement. *Clin. Chem.* 1992. 38:1632-1638.

[44] Jeyarajah EJ, Cromwell WC, Otvos JD. Lipoprotein particle analysis by nuclear magnetic resonance spectroscopy. *Clin. Lab Med.* 2006. 26:847-870.

[45] El Harchaoui K, Arsenault BJ, Franssen R, Després JP, Hovingh GK, Stroes ES, Otvos JD, Wareham NJ, Kastelein JJ, Khaw KT, Boekholdt SM. High-density lipoprotein particle size and concentration and coronary risk. *Ann. Intern Med.* 2009. 150:84-93.

[46] Mora S, Szklo M, Otvos JD, Greenland P, Psaty BM, Goff DC Jr, O'Leary DH, Saad MF, Tsai MY, Sharrett AR. LDL particle subclasses, LDL particle size, and carotid atherosclerosis in the Multi-Ethnic Study of Atherosclerosis (MESA). *Atherosclerosis.* 2007. 192:211-217.

[47] Otvos JD, Collins D, Freedman DS, Shalaurova I, Schaefer EJ, McNamara JR, Bloomfield HE, Robins SJ. Low-density lipoprotein and high-density lipoprotein particle subclasses predict coronary events and are favorably changed by gemfibrozil therapy in the Veterans Affairs High-Density Lipoprotein Intervention Trial. *Circulation.* 2006. 113:1556-1563.

[48] Rosenson RS, Otvos JD, Freedman DS. Relations of lipoprotein subclass levels and low-density lipoprotein size to progression of coronary artery

disease in the Pravastatin Limitation of Atherosclerosis in the Coronary Arteries (PLAC-I) trial. *Am. J. Cardiol.* 2002. 90:89-94.

[49] Ridker PM, Genest J, Boekholdt SM, Libby P, Gotto AM, Nordestgaard BG, Mora S, MacFadyen JG, Glynn RJ, Kastelein JJ; JUPITER Trial Study Group. HDL cholesterol and residual risk of first cardiovascular events after treatment with potent statin therapy: an analysis from the JUPITER trial. *Lancet.* 2010. 376:333-339.

[50] Drexel H, Aczel S, Marte T, Vonbank A, Saely CH. Factors predicting cardiovascular events in statin-treated diabetic and non-diabetic patients with coronary atherosclerosis. *Atherosclerosis.* 2010. 208:484-489.

[51] Nicholls SJ, Tuzcu EM, Sipahi I, Grasso AW, Schoenhagen P, Hu T, Wolski K, Crowe T, Desai MY, Hazen SL, Kapadia SR, Nissen SE. Statins, high-density lipoprotein cholesterol, and regression of coronary atherosclerosis. *JAMA.* 2007. 297:499-508.

[52] Felix-Getzik EM, Kuvin JT, Karas RH. Nonoptimal high-density lipoprotein cholesterol levels are highly prevalent in patients presenting with acute coronary syndromes and well-controlled low-density lipoprotein cholesterol levels. *J. Clin. Lipidol.* 2010. 4:265-271.

[53] Saely CH, Vonbank A, Drexel H. HDL cholesterol and residual risk of first cardiovascular events. *Lancet.* 2010. 376:1738; author reply 1738-1739.

[54] Zeljkovic A, Vekic J, Spasojevic-Kalimanovska V, Jelic-Ivanovic Z, Bogavac-Stanojevic N, Gulan B, Spasic S. LDL and HDL subclasses in acute ischemic stroke: prediction of risk and short-term mortality. *Atherosclerosis.* 2010. 210:548-554.

[55] El Harchaoui K, Arsenault BJ, Franssen R, Després JP, Hovingh GK, Stroes ES, Otvos JD, Wareham NJ, Kastelein JJ, Khaw KT, Boekholdt SM. High-density lipoprotein particle size and concentration and coronary risk. *Ann. Intern. Med.* 2009. 150:84-93.

[56] Caulfield MP, Li S, Lee G, Blanche PJ, Salameh WA, Benner WH, Reitz RE, Krauss RM. Direct determination of lipoprotein particle sizes and concentrations by ion mobility analysis. *Clin. Chem.* 2008. 54:1307-1316.

[57] Musunuru K, Orho-Melander M, Caulfield MP, Li S, Salameh WA, Reitz RE, Berglund G, Hedblad B, Engström G, Williams PT, Kathiresan S, Melander O, Krauss RM. Ion mobility analysis of lipoprotein subfractions identifies three independent axes of cardiovascular risk. *Arterioscler. Thromb Vasc Biol.* 2009 29:1975-1980.

[58] Rosenson RS, Brewer HB Jr, Chapman MJ, Fazio S, Hussain MM, Kontush A, Krauss RM, Otvos JD, Remaley AT, Schaefer EJ. HDL measures, particle heterogeneity, proposed nomenclature, and relation to atherosclerotic cardiovascular events. *Clin Chem*. 2011. 57:392-410.

[59] Alaupovic P. The concept of apolipoprotein-defined lipoprotein families and its clinical significance. *Curr. Atheroscler Rep*. 2003. 5:459-467.

[60] *Parra HJ, Mezdour H, Ghalim N, Bard JM, Fruchart JC. Differential electroimmunoassay of human LpA-I lipoprotein particles on ready-to-use plates. Clin. Chem.* 1990. 36*:* 1431*–1435.*

[61] Puchois P, Kandoussi A, Fievet P, Fourrier JL, Bertrand M, Koren E, Fruchart JC. Apolipoprotein A-I containing lipoproteins in coronary artery disease. *Atherosclerosis*. 1987. 68:35-40.

[62] Parra HJ, Arveiler D, Evans AE, Cambou JP, Amouyel P, Bingham A, McMaster D, Schaffer P, Douste-Blazy P, Luc G, et al. A case-control study of lipoprotein particles in two populations at contrasting risk for coronary heart disease. The ECTIM Study. *Arterioscler. Thromb*. 1992. 12:701-707.

[63] van der Steeg WA, Holme I, Boekholdt SM, Larsen ML, Lindahl C, Stroes ES, Tikkanen MJ, Wareham NJ, Faergeman O, Olsson AG, Pedersen TR, Khaw KT, Kastelein JJ. High-density lipoprotein cholesterol, high-density lipoprotein particle size, and apolipoprotein A-I: significance for cardiovascular risk: the IDEAL and EPIC-Norfolk studies. *J. Am. Coll Cardiol*. 2008. 51:634-642.

[64] Walldius G, Jungner I, Holme I, Aastveit AH, Kolar W, Steiner E. High apolipoprotein B, low apolipoprotein A-I, and improvement in the prediction of fatal myocardial infarction (AMORIS study): a prospective study. Lancet. 2001. 358:2026-2033.

[65] Ferretti G, Bacchetti T, Nègre-Salvayre A, Salvayre R, Dousset N, Curatola G. Structural modifications of HDL and functional consequences. *Atherosclerosis*. 2006. 184:1-7.

[66] Zheng L, Nukuna B, Brennan ML, Sun M, Goormastic M, Settle M, Schmitt D, Fu X, Thomson L, Fox PL, Ischiropoulos H, Smith JD, Kinter M, Hazen SL. Apolipoprotein A-I is a selective target for myeloperoxidase-catalyzed oxidation and functional impairment in subjects with cardiovascular disease. *J. Clin. Invest*. 2004. 114:529-541.

[67] Birjmohun RS, Dallinga-Thie GM, Kuivenhoven JA, Stroes ES, Otvos JD, Wareham NJ, Luben R, Kastelein JJ, Khaw KT, Boekholdt SM. Apolipoprotein A-II is inversely associated with risk of future coronary artery disease. *Circulation*. 2007. 116:2029-2035.

[68] Winkler K, Hoffmann MM, Seelhorst U, Wellnitz B, Boehm BO, Winkelmann BR, März W, Scharnagl H. Apolipoprotein A-II Is a negative risk indicator for cardiovascular and total mortality: findings from the Ludwigshafen Risk and Cardiovascular Health Study. *Clin. Chem.* 2008. 54:1405-1406.
[69] Alaupovic P, Mack WJ, Knight-Gibson C, Hodis HN. The role of triglyceride-rich lipoprotein families in the progression of atherosclerotic lesions as determined by sequential coronary angiography from a controlled clinical trial. *Arterioscler Thromb. Vasc. Biol.* 1997. 17*: 715–722.*
[70] Warden CH, Hedrick CC, Qiao JH, Castellani LW, Lusis AJ. Atherosclerosis in transgenic mice overexpressing apolipoprotein A-II. *Science*. 1993. 261:469-72.
[71] de la Llera Moya M, Atger V, Paul JL, Fournier N, Moatti N, Giral P, Friday KE, Rothblat G. A cell culture system for screening human serum for ability to promote cellular cholesterol efflux. Relations between serum components and efflux, esterification, and transfer. *Arterioscler Thromb.* 1994. 14:1056-1065.
[72] Fielding PE, Fielding CJ, Havel RJ, Kane JP, Tun P. Cholesterol net transport, esterification, and transfer in human hyperlipidemic plasma. *J. Clin. Invest.* 1983. 71:449-460.
[73] Singaraja RR, Fievet C, Castro G, James ER, Hennuyer N, Clee SM, Bissada N, Choy JC, Fruchart JC, McManus BM, Staels B, Hayden MR. Increased ABCA1 activity protects against atherosclerosis. *J. Clin. Invest.* 2002. 110:35-42.
[74] van Dam MJ, de Groot E, Clee SM, Hovingh GK, Roelants R, Brooks-Wilson A, Zwinderman AH, Smit AJ, Smelt AH, Groen AK, Hayden MR, Kastelein JJ. Association between increased arterial-wall thickness and impairment in ABCA1-driven cholesterol efflux: an observational study. *Lancet*. 2002. 359:37-42.
[75] Navab M, Anantharamaiah GM, Reddy ST, Hama S, Hough G, Grijalva VR, Wagner AC, Frank JS, Datta G, Garber D, Fogelman AM. Oral D-4F causes formation of pre-beta high-density lipoprotein and improves high-density lipoprotein-mediated cholesterol efflux and reverse cholesterol transport from macrophages in apolipoprotein E-null mice. *Circulation*. 2004. 109:3215-3220.
[76] Yvan-Charvet L, Matsuura F, Wang N, Bamberger MJ, Nguyen T, Rinninger F, Jiang XC, Shear CL, Tall AR. Inhibition of cholesteryl ester transfer protein by torcetrapib modestly increases macrophage

cholesterol efflux to HDL. *Arterioscler Thromb. Vasc. Biol.* 2007 May. 27:1132-1138.

[77] Eriksson M, Carlson LA, Miettinen TA, Angelin B. Stimulation of fecal steroid excretion after infusion of recombinant proapolipoprotein A-I. Potential reverse cholesterol transport in humans. *Circulation.* 1999. 100:594-598.

[78] Voogt J, Killion S, Murphy L, Hellerstein MK, Turner SM. Measurement of cholesterol efflux and global cholesterol transport rates in vivo with stable isotopes. *Paper presented at: Arteriosclerosis, Thrombosis, and Vascular Biology Annual Conference*; April 19–21, 2007; Chicago, IL.

[79] Barter PJ, Nicholls S, Rye KA, Anantharamaiah GM, Navab M, Fogelman AM. Antiinflammatory properties of HDL. *Circ. Res.* 2004. 95:764-772.

[80] Navab M, Hough GP, Stevenson LW, Drinkwater DC, Laks H, Fogelman AM. Monocyte migration into the subendothelial space of a coculture of adult human aortic endothelial and smooth muscle cells. *J. Clin. Invest.* 1988. 82:1853-1863.

[81] Ansell BJ, Navab M, Hama S, Kamranpour N, Fonarow G, Hough G, Rahmani S, Mottahedeh R, Dave R, Reddy ST, Fogelman AM. Inflammatory/antiinflammatory properties of high-density lipoprotein distinguish patients from control subjects better than high-density lipoprotein cholesterol levels and are favorably affected by simvastatin treatment. *Circulation.* 2003. 108:2751-2756.

[82] Navab M, Hama SY, Anantharamaiah GM, Hassan K, Hough GP, Watson AD, Reddy ST, Sevanian A, Fonarow GC, Fogelman AM. Normal high density lipoprotein inhibits three steps in the formation of mildly oxidized low density lipoprotein: steps 2 and 3. *J. Lipid Res.* 2000. 41:1495-1508.

[83] Navab M, Hama SY, Hough GP, Subbanagounder G, Reddy ST, Fogelman AM. A cell-free assay for detecting HDL that is dysfunctional in preventing the formation of or inactivating oxidized phospholipids. *J. Lipid Res.* 2001. 42:1308-1317.

[84] Dodani S, Kaur R, Reddy S, Reed GL, Navab M, George V. Can dysfunctional HDL explain high coronary artery disease risk in South Asians? *Int J Cardiol.* 2008. 129:125-132.

[85] Lawrence MB, Springer TA. Leukocytes roll on a selectin at physiologic flow rates: distinction from and prerequisite for adhesion through integrins. *Cell.* 1991. 65:859-873.

[86] Davies MJ, Gordon JL, Gearing AJ, Pigott R, Woolf N, Katz D, Kyriakopoulos A. The expression of the adhesion molecules ICAM-1, VCAM-1, PECAM, and E-selectin in human atherosclerosis. *J. Pathol.* 1993. 171:223-229.
[87] Kume N, Cybulsky MI, Gimbrone MA Jr. Lysophosphatidylcholine, a component of atherogenic lipoproteins, induces mononuclear leukocyte adhesion molecules in cultured human and rabbit arterial endothelial cells. *J. Clin. Invest.* 1992. 90:1138-1144.
[88] Krejcy K, Schwarzacher S, Ferber W, Plesch C, Cybulsky MI, Weidinger FF. Expression of VCAM-1 in rabbit iliac arteries is associated with vasodilator dysfunction of regenerated endothelium following balloon injury. *Atherosclerosis*. 1996. 122:59-67.
[89] Iiyama K, Hajra L, Iiyama M, Li H, DiChiara M, Medoff BD, Cybulsky MI. Patterns of vascular cell adhesion molecule-1 and intercellular adhesion molecule-1 expression in rabbit and mouse atherosclerotic lesions and at sites predisposed to lesion formation. *Circ. Res.* 1999. 85:199-207.
[90] Rohde LE, Lee RT, Rivero J, Jamacochian M, Arroyo LH, Briggs W, Rifai N, Libby P, Creager MA, Ridker PM. Circulating cell adhesion molecules are correlated with ultrasound-based assessment of carotid atherosclerosis. *Arterioscler Thromb. Vasc. Biol.* 1998. 18:1765-1770.
[91] Ridker PM, Hennekens CH, Roitman-Johnson B, Stampfer MJ, Allen J. Plasma concentration of soluble intercellular adhesion molecule 1 and risks of future myocardial infarction in apparently healthy men. *Lancet.* 1998. 351:88-92.
[92] Barter PJ, Baker PW, Rye KA. Effect of high-density lipoproteins on the expression of adhesion molecules in endothelial cells. *Curr. Opin Lipidol.* 2002. 13:285-288.
[93] Cockerill GW, Rye KA, Gamble JR, Vadas MA, Barter PJ. High-density lipoproteins inhibit cytokine-induced expression of endothelial cell adhesion molecules. *Arterioscler Thromb Vasc. Biol.* 1995. 15:1987-1994.
[94] Ashby DT, Rye KA, Clay MA, Vadas MA, Gamble JR, Barter PJ. Factors influencing the ability of HDL to inhibit expression of vascular cell adhesion molecule-1 in endothelial cells. *Arterioscler Thromb Vasc. Biol.* 1998. 18:1450-1455.
[95] Baker PW, Rye KA, Gamble JR, Vadas MA, Barter PJ. Phospholipid composition of reconstituted high density lipoproteins influences their

ability to inhibit endothelial cell adhesion molecule expression. *J. Lipid Res*. 2000. 41:1261-1267.

[96] Oguchi S, Dimayuga P, Zhu J, Chyu KY, Yano J, Shah PK, Nilsson J, Cercek B. Monoclonal antibody against vascular cell adhesion molecule-1 inhibits neointimal formation after periadventitial carotid artery injury in genetically hypercholesterolemic mice. *Arterioscler Thromb. Vasc. Biol*. 2000. 20:1729-1736.

[97] Uittenbogaard A, Shaul PW, Yuhanna IS, Blair A, Smart EJ. High density lipoprotein prevents oxidized low density lipoprotein-induced inhibition of endothelial nitric-oxide synthase localization and activation in caveolae. *J. Biol. Chem*. 2000. 275:11278–11283.

[98] Braman RS, Hendrix SA. Nanogram nitrite and nitrate determination in environmental and biological materials by vanadium (III) reduction with chemiluminescence detection. *Anal. Chem*. 1989. 61: 2715–2718.

[99] Ou Z, Ou J, Ackerman AW, Oldham KT, Pritchard KA Jr. L-4F, an apolipoprotein A-1 mimetic, restores nitric oxide and superoxide anion balance in low-density lipoprotein-treated endothelial cells. *Circulation*. 2003. 107:1520–1524.

[100] Ou J, Wang J, Xu H, et al. Effects of D-4F on vasodilation and vessel wall thickness in hypercholesterolemic LDL receptor-null and LDL receptor/apolipoprotein A-I double-knockout mice on Western diet. *Circ. Res*. 2005. 97:1190 –1197.

[101] Shaul PW. Endothelial nitric oxide synthase, caveolae and the development of atherosclerosis. *J. Physiol*. 2003. 547:21–33.

[102] Kimura T, Sato K, Malchinkhuu E, et al. High-density lipoprotein stimulates endothelial cell migration and survival through sphingosine 1-phosphate and its receptors. *Arterioscler Thromb. Vasc* Biol. 2003. 23:1283–1288.

[103] Nofer JR, Levkau B, Wolinska I, et al. Suppression of endothelial cell apoptosis by high density lipoproteins (HDL) and HDL-associated lysosphingolipids. *J. Biol. Chem*. 2001. 276:34480 –34485.

[104] Kuvin JT, Ramet ME, Patel AR, Pandian NG, Mendelsohn ME, Karas RH. A novel mechanism for the beneficial vascular effects of high-density lipoprotein cholesterol: enhanced vasorelaxation and increased endothelial nitric oxide synthase expression. *Am. Heart J*. 2002. 144:165–172.

[105] Li XP, Zhao SP, Zhang XY, Liu L, Gao M, Zhou QC. Protective effect of high density lipoprotein on endothelium-dependent vasodilatation. *Int J. Cardiol*. 2000. 73:231– 236.

[106] Zhang X, Zhao SP, Li XP, Gao M, Zhou QC. Endotheliumdependent and -independent functions are impaired in patients with coronary heart disease. *Atherosclerosis*. 2000. 149:19 –24.
[107] Kuvin JT, Patel AR, Sidhu M, et al. Relation between high-density lipoprotein cholesterol and peripheral vasomotor function. *Am. J. Cardiol*. 2003. 92:275–279.
[108] Lupattelli G, Marchesi S, Roscini AR, et al. Direct association between high-density lipoprotein cholesterol and endothelial function in hyperlipemia. *Am. J. Cardiol*. 2002. 90:648 –650.
[109] Toikka JO, Ahotupa M, Viikari JS, et al. Constantly low HDL cholesterol concentration relates to endothelial dysfunction and increased in vivo LDL-oxidation in healthy young men. *Atherosclerosis*. 1999. 147:133– 138.
[110] Wang TD, Chen WJ, Lin JW, Cheng CC, Chen MF, Lee YT. Efficacy of fenofibrate and simvastatin on endothelial function and inflammatory markers in patients with combined hyperlipidemia: relations with baseline lipid profiles. *Atherosclerosis*. 2003. 170:315–323.
[111] Fathi R, Haluska B, Isbel N, Short L, Marwick TH. The relative importance of vascular structure and function in predicting cardiovascular events. *J. Am. Coll Cardiol*. 2004. 43:616 –623.
[112] Li D, Weng S, Yang B, et al. Inhibition of arterial thrombus formation by ApoA1 Milano. *Arterioscler Thromb Vasc. Biol.* 1999. 19:378–383.
[113] Naqvi TZ, Shah PK, Ivey PA, et al. Evidence that high-density lipoprotein cholesterol is an independent predictor of acute platelet dependent thrombus formation. *Am J Cardiol*. 1999. 84:1011–1017.
[114] Eitzman DT, Westrick RJ, Shen Y, et al. Homozygosity for factor V Leiden leads to enhanced thrombosis and atherosclerosis in mice. *Circulation*. 2005. 111:1822–1825.
[115] Griffin JH, Fernandez JA, Deguchi H. Plasma lipoproteins, hemostasis and thrombosis. *Thromb. Haemost*. 2001. 86:386 –394.
[116] Griffin JH, Kojima K, Banka CL, Curtiss LK, Fernandez JA. High-density lipoprotein enhancement of anticoagulant activities of plasma protein S and activated protein C. *J. Clin. Invest*. 1999. 103: 219–227.
[117] Carson SD. Plasma high density lipoproteins inhibit the activation of coagulation factor X by factor VIIa and tissue factor. *FEBS Lett*. 1981. 132:37– 40.
[118] Levin EG, Miles LA, Fless GM, et al. Lipoproteins inhibit the secretion of tissue plasminogen activator from humanendothelial cells. *Arterioscler Thromb*. 1994. 14:438–442.

[119] Viswambharan H, Ming XF, Zhu S, et al. Reconstituted high density lipoprotein inhibits thrombin-induced endothelial tissue factor expression through inhibition of RhoA and stimulation of phosphatidylinositol 3-kinase but not Akt/endothelial nitric oxide synthase. *Circ. Res*. 2004. 94:918 –925.

[120] Chesterman CN, Berndt MC. Platelet and vessel wall interaction and the genesis of atherosclerosis. *Clin. Haematol*. 1986. 15:323–353.

[121] Huo Y, Schober A, Forlow SB, et al. Circulating activated platelets exacerbate atherosclerosis in mice deficient in apolipoprotein E. *Nat. Med*. 2003. 9:61–67.

[122] Massberg S, Brand K, Gruner S, et al. A critical role of platelet adhesion in the initiation of atherosclerotic lesion formation. *J. Exp. Med*. 2002. 196:887–896.

In: Dyslipidemia
Editors: Z. Ačimovič et.al, pp. 87-114
ISBN: 978-1-62100-917-7

Chapter 3

DYSLIPIDEMIA: CAUSES, DIAGNOSIS AND TREATMENT

Athanasia K. Papazafiropoulou*, Marina S. Kardara and Stavros I. Pappas

3rd Department of Internal Medicine and Center of Diabetes,
General Hospital of Nikaia "Ag. Panteleimon" – Piraeus, Greece

ABSTRACT

Cardiovascular disease (CVD) is leading cause of morbidity and worldwide. Higher levels of low-density lipoprotein cholesterol (LDL-C) are associated with an increased risk of coronary heart disease (CHD), myocardial infarction, and stroke. Guidelines of the Adult Treatment Panel III emphasize intensive reduction of LDL or non-high-density lipoprotein cholesterol in patients at high risk of CHD.

Restriction of dietary saturated and trans-fat and cholesterol, along with increased intake of soluble fiber, can achieve LDL-C lowering. However, pharmacological treatment is needed in most patients. Statins are highly efficacious as LDL-C lowering agents and have more modest effects on very LDL-C, triglycerides and HDL-C levels. The effects of statins on plasma lipids result from their ability to both increase the

* Correspondence: A. K. Papazafiropoulou, MD, 3rd Department of Internal Medicine and Center of Diabetes, General Hospital of Nikaia "Ag. Panteleimon" – Piraeus, Greece, 3 D.Mantouvalou Street, ,R-184 54 Nikaia, Greece, Tel : +30 213 207 2662, Fax : +30 213 207 6404, E-mail: pathan@ath.forthnet.gr

efficiency with which very LDL-C and LDL-C are cleared from the circulation and reduce the production of apoB-containing lipoproteins by the liver. Placebo-controlled intervention studies of statin drugs for lowering LDL-C provide clear evidence of CVD prevention. In addition, improvement of survival with statins may be due to other pleiotropic effects beyond LDL-C lowering. Side-effects, myopathy and hepatotoxicity, are infrequent and usually mild, but use of lipid-modifying medication demands caution because of the possibility drug interactions.

Many recent studies suggest that low levels of HDL-C are a major independent risk factor for CVD. According to several clinical trials, a 1% increase in HDL-C is associated with a 0.7%–3% decrease in CHD events. The direct link between high levels of triglycerides and CHD is less well defined. Fibrates reduce triglycerides levels and may be beneficial for CHD prevention. Fibric acid derivatives and niacin are primarily used to increase HDL-C levels, although with side effects. Low-dose n-3 fatty acids could be used routinely after a myocardial infarction, but the value of higher doses of n-3 fatty acids in reducing cardiovascular risk remains to be demonstrated.

In conclusion, achieving low lipid levels appears safe, but the generalizability of these findings to broader populations and the clinical benefit on the reduction of cardiovascular complications remains to be proven.

INTRODUCTION

It is well known that coronary heart disease (CHD) is a leading cause of morbidity and mortality worldwide [1]. Only in the United States, one of every five deaths was attributed to CHD in 2002 [1]. Therefore, CHD has a major impact in health economics with the estimated total cost for CHD in 2005 exceeded $142 billion [1].

The National Cholesterol Education Program Adult Treatment Panel's third report (NCEP ATP-III) focuses on evidence from clinical trials demonstrating the importance of LDL-C reduction to reduce the risk of CHD [2].

Lowering LDL-C to less than 100 mg/dL is recommended for those with known CHD or CHD risk equivalents such as diabetes mellitus (DM). For subjects at high-risk, a target LDL-C of < 70 mg/dL has been added aiming to the benefits of aggressive lipid-lowering.

Cholesterol and Lipoprotein Metabolism and Transport

The major lipids are cholesterol, triglycerides (TGs), and phospholipids, which are used for cell membrane formation, hormone synthesis, and bile acid production [3, 4]. Since lipids are not soluble in blood, they are packaged into complexes of lipid and protein called lipoprotein particles [3]. Lipoproteins are composed of phospholipids, free cholesterols, proteins (Apo lipoproteins), cholesterol esters, and TGs [3]. There are 5 major classes of lipoprotein particles: chylomicrons, very low-density lipoprotein cholesterol (VLDL-C), intermediate density lipoproteins, LDL-C, and high-density lipoprotein cholesterol (HDL-C) [3].

Chylomicrons are the largest lipoprotein particles, and transport dietary TGs and cholesterols from the intestine to different parts of the body [3, 5]. The metabolism of chylomicrons is mediated by lipoprotein lipase (LPL), with ApoC-II acting as a cofactor and activator [3, 5]. Chylomicrons are usually present in the plasma for 3–6 hours after eating and are metabolized after 10 to 12 hours of fasting [5]. After LPL metabolizes the chylomicrons and removes a large portion of the TGs, chylomicrons are further metabolized in the liver by hepatic lipase, returning ApoC-II to HDL-C [3, 5]. Chylomicron remnants, which contain Apo-E and ApoB-4, are then taken up by remnant receptors in the liver, which liberates free cholesterols intracellularly [3]. During the whole process of chylomicron metabolism, some of the components of the chylomicron such as phospholipids and some Apo lipoproteins are recycled and used to make HDL-C [3].

VLDL-C is synthesized in the liver in response to a high carbohydrate diet. Excess carbohydrate is converted into TGs and transferred to nascent VLDL-C. VLDL-C delivers TGs to adipose tissue and cardiac or skeletal muscle for storage and energy release by LPL. Almost all LDL-C is derived from VLDL-C [5].

LDL-C carries about 60% to 70% of serum cholesterol [3]. It transports cholesterol from the liver to peripheral tissues. High levels of LDL-C are harmful, as LDL-C can build up on arterial walls, leading to the formation of atherosclerotic plaques. The binding of LDL-C to its receptor in the liver is the major mechanism of removal of LDL-C from the circulation [6]. An increase in intracellular cholesterol inhibits *de novo* synthesis of cholesterol, resulting in decreased synthesis of LDL-C receptor and increased activity of an enzyme that facilitates cholesterol storage [3].

HDL-C is known to be protective against CHD, while low HDL-C levels increase the risk of CHD. HDL-C is produced in the intestine, liver and plasma as a complex of Apo A lipoproteins, phospholipids and cholesterol. In the plasma, HDL-C is converted to a cholesterol ester by the action of cholesterol ester transferase (LCAT). As they circulate in the blood stream, HDL-C particles acquire more cholesterol from the blood stream. In addition, HDL-C particles remove cholesterol through a reverse cholesterol transport process from peripheral tissues and atheroma within the arteries to the liver, carrying approximately 30% of the serum cholesterol [5]. Women and individuals on estrogen therapy generally have higher HDL-C levels due to the effect of estrogen on the upregulation of the major components of the HDL-C particle, mainly ApoA-1 lipoprotein [5].

The process known as reverse cholesterol transport involves removal by HDL-C of un-esterified cholesterol from peripheral cells, such as macrophages, and delivery to the liver through the interaction of HDL-C with the hepatic HDL-C receptor. Several studies have shown an inverse relationship between blood HDL-C levels and CHD. These studies show that individuals with low levels of HDL-C have worse CHD outcomes. Furthermore, data obtained from several epidemiological studies emphasize that the risk factor of low HDL-C is completely independent of LDL-C; no matter how low the LDL-C, a decrease in the HDL-C would increase the risk for coronary artery disease [3].

Pathogenesis of Dyslipidemia

There are many causes of dyslipidemia, both primary and secondary. The primary causes of dyslipidemia are due mostly to genetic disorders. There are six categories of lipoprotein disorders according to the Fredrickson-Levy-Lees classification: Type I (high levels of chylomicrons), Type IIa (high LDL-C levels), Type IIb (high LDL-C and VLDL-C levels), Type III (high IDL-C levels), Type IV (high VLDL-C levels), and Type V (high LDL-C and chylomicrons levels) [3]. It is also possible that some disease states can be placed into more than one category of lipoprotein disorder.

The term hypertriglyceridemia usually refers to elevations in VLDL-C and chylomicrons, both of which carry and transport TGs [3]. Primary hypertriglyceridemia includes primary chylomicronemia, familial hypertriglyceridemia, familial combined hyperlipoproteinemia and familial dysbetalipoproteinsemia [6]. Primary chylomicronemia is a genetic disease that is characterized by a deficiency in LPL or cofactor, and results in elevated chylomicrons and VLDL-C and severe elevation of TGs, leading to acute pancreatitis [6]. Familial hypertriglyceridemia is a Type IV disorder in which primarily VLDL-C is affected [3]. It is caused by a number of genetic determinants that result in insufficient removal of TG-rich lipoproteins [6]. Familial combined hyperlipoproteinemia is a disorder characterized by increased levels of VLDL-C, LDL-C, or both. Familial dysbeta lipoproteinemia is a disorder characterized by increased levels of VLDL-C remnant and chylomicron remnant [3, 6].

The term hypercholesterolemia usually refers to elevated serum LDL-C. Primary hypercholesterolemia includes familial hypercholesterolemia, familial ligand-defective Apo lipoprotein B, and familial combined hyperlipoproteinemia. Primary hypercholesterolemia is a Type IIa dominant disorder that involves mutations in the LDL-C receptor gene. Homozygotes usually have a worse prognosis than heterozygotes. Familial ligand-defective apoliporotein B is also a Type IIa disorder caused by a mutation in ApoB-100 that disrupts the binding of LDL-C to the LDL-C receptor, thereby decreasing metabolism of LDL-C. In both of these disorders, LDL-C receptor-mediated endocytosis in the liver is decreased, resulting in increased serum LDL-C. In addition, genetic evidence confirms the role of a newly discovered serine protease, pro-protein convertase subtilisin-like kexin type 9 (PCSK9), in patients suffering from autosomal dominant hypercholesterolemia. Three gain of function single nucleotide polymorphisms in the PCSK9 gene have been identified that increase LDL-C levels. PCSK9 destroys low density lipoproteins receptors (LDL-C-R) in the liver, thereby regulating the levels of LDL-C in plasma. Lastly, familial combined hyperlipoproteinemia is a Type IIb disorder that can cause elevated VLDL-C, LDL-C, or both.

Some rare genetic disorders can also cause low levels of HDL-C, ie, Tangier disease, disorders of LCAT and familial hypoalphalipoproteinemia. Tangier disease is caused by mutation and loss of function of ATP-binding cassette 1 (ABCA1). ABCA1 transports cholesterol and phospholipids out of cells for pickup by Apo-A1 in the circulation. In the absence of ABCA1, free Apo-A1 does not acquire cellular lipids. This results in accelerated clearance of Apo-A1 from plasma, leading to low HDL-C levels [7]. LCAT is a lecithin cholesterol acyltransferase responsible for catalyzing the formation of cholesterol esters of HDL-C and LDL-C; thus, it is crucial for HDL-C formation [8]. Familial hypoalphalipoproteinemia includes a wide range of disorders that result in low levels of HDL-C (usually below 35 mg/dL).

There are many secondary causes of dyslipidemia, including disease- and medication-induced dyslipidemia. Some of the more common causes of hypertriglyceridemia are DM, alcohol abuse, estrogen, and obesity [3, 6]. Some of the more common causes of hypercholesterolemia are hypothyroidism, anorexia, and excess corticosteroid use [3, 6]. Low levels of HDL-C can be caused by malnutrition, obesity, and drugs such as beta-blockers, anabolic steroids, isotretinoin, and progestins [3].

CHD and Dyslipidemia

High levels of LDL-C (and to a lesser extent, VLDL-C) result in the accumulation of LDL-C in the arterial wall, leading to oxidation of LDL-C [3]. Oxidized LDL-C can cause extensive damage to the arterial wall, provoking inflammation responses, promoting coagulation, increasing the activity of mediators that cause vasoconstriction and inhibiting mediators that cause vasodilation [3]. Oxidized LDL-C recruits monocytes, which enter the arterial wall and are activated to become macrophages. The macrophages ingest oxidized LDL-C through the macrophage scavenger receptor to become foam cells, or fatty streak. Foam cells propagate inflammatory responses as well as facilitate deposition of more oxidized LDL-C. Micro-calcification of the vascular smooth muscle cells will take place, which progresses to atherosclerosis [3].

The fatty streak, which consists of cholesterol-filled macrophages, is the first stage of atherosclerosis [3]. Plaques (deposits of fatty substances, cholesterol, calcium, and cell components) will then form and progress, gradually increasing inside the artery and narrowing the arterial wall, resulting in decreased flow of blood and oxygen to tissues. Plaques are usually kept in check by a fibrous cap, which protects and stabilizes the lesion. If the plaque ruptures, then thrombosis will occur and damage will spread to other areas, leading to ischemic heart disease, myocardial infraction, stroke, peripheral arterial disease, other CHDs, and possibly death [3].

HDL-C is believed to help protect against CHD. Its protective effects are in part due to reverse cholesterol transport. As the name suggests, the reverse cholesterol transport process involves the transport of HDL-C from peripheral tissues and transfer to VLDL-C and LDL-C back to liver for secretion in the bile [3]. HDL-C is also believed to have anti-inflammatory, anti-oxidative, platelet anti-aggregatory, anticoagulation, and pro-fibrolytic effects, which can help reduce damage mediated by high levels of LDL-C [7].

Treatment of Dyslipidemia

New classifications of LDL-C, total cholesterol (TC), and HDL-C levels are identified in Table 2. In addition, ATP III defined the desirable TGs level as 150 mg/dL and set non–HDL-C treatment goals in patients with TGs >200 mg/dL.

The NCEP ATP III guidelines stratify patients into 1 of 3 risk categories according to their near-term risk of developing CHD [9]. The first risk category includes patients with preexisting CHD or CHD risk equivalents and patients with a very high risk, defined as a greater than 20% risk for developing CHD in the next 10 years. The second category consists of patients with 2 or more risk factors but not diagnosed with CHD, and the third category includes those who are in the lowest risk category, with 0 to 1 risk factors (Table 1). Treatment strategies are driven by the degree of risk for CHD; the higher the risk, the more aggressive the treatment strategy.

ATP III recommends the use of a Framingham based risk scoring system to quantify the 10-year risk for a coronary event [9]. Point scores are calculated according to the presence of 5 major CHD risk factors—age, gender, TC level, systolic blood pressure, HDL-C, and smoking status—with each risk factor worth a certain number of points. When added together, the sum yields an estimate of the 10-year risk for experiencing a coronary event. A properly conducted assessment places patients into 1 of the 3 risk categories and forms the basis for all subsequent treatment decisions [9]. Patients with documented CHD and CHD risk equivalents are automatically placed in the highest risk category. CHD risk equivalents carry a risk for a major coronary event equal to that of clinically evident CHD and include type 2 diabetes (T2D), peripheral vascular disease, symptomatic carotid artery disease, and abdominal aortic aneurysm [9].

The LDL-C treatment goal for patients in this high-risk category is a level <100 mg/dL. The first step in evaluating CHD risk is to identify patients with clinical evidence of CHD or CHD risk equivalents. All of these patients, by definition, have a >20% 10-year risk of a CHD event. The second step in assessing risk is to count the risk factors present in the remaining patients (ie, those without CHD or CHD risk equivalents). For those with >2 risk factors, the next step is to conduct a Framingham-based risk assessment to determine the most appropriate course of therapy. If the patient presents ≥2 risk factors, he/she may have a 10-year CHD risk which is >20%. If this is the case, this patient should be placed in the CHD risk equivalent category and treated to a target LDL-C goal of <100 mg/dL [9].

Table 1. Risk Assessment Criteria: New Features of NCEP ATP III Guidelines

Focus on Multiple Risk Factors

- Diabetes without CHD equal to CHD risk equivalent
- Framingham-based assessment of 10-year absolute CHD risk used to stratify patients with >2 risk factors into those with >20% risk for more intensive treatment and those with <20% risk
- Multiple metabolic risk factors indicate candidates for intensified therapeutic lifestyle changes

Modifications of Lipid and Lipoprotein Classification

- LDL-C <100 mg/dL optimal
- Raises categorical low HDL-C from <35 mg/dL to 40 mg/dL
- Lowers TG classification cutpoints to give more attention to moderate elevations

Support for Implementation

- Lipoprotein analysis versus screening for TC and HDL-C alone as preferred initial test
- Encourage dietary options (plant stanols/sterols and soluble fiber) to enhance lowering LDL-C
- Promote adherence to therapeutic lifestyle changes and drug therapies
- Treat to a second, non–HDL-C treatment goal in patients who have a TG ≥200 mg/dL

ATP III indicates Adult Treatment Panel III; HDL-C, high-density lipoprotein cholesterol; LDL-C, low-density lipoprotein cholesterol; NCEP, National
Cholesterol Education Program; TC, total cholesterol; and TG, triglycerides.

Table 2. NCEP ATP III Classification of Cholesterol: Total, LDL-C, HDL-C

Total Cholesterol (mg/dL)

- <200 Desirable
- 200-239 Borderline
- ≥240 High

LDL-C (mg/dL)

- <100 Optimal
- 100-129 Near optimal/above optimal
- 130-159 Borderline high
- 160-189 High
- ≥190 Very high

HDL-C (mg/dL)

- <40 Low
- ≥60 High

Patients with a 10-year risk between 10% and 20% require more aggressive treatment, often with lifestyle modifications and drug therapy. Therapy for these patients should be sufficient to enable patients in this category to achieve an LDL-C target of <130 mg/dL. Those patients with <10% 10-year risk are candidates for lifestyle therapy and rarely, drug therapy. The target LDL-C in this group of patients is <160 mg/dL [9].

Therapeutic Lifestyle Changes

ATP III continues to stress the importance of non pharmacologic treatment but recognizes its limitations by reducing the trial of these modalities from six months to 12 weeks before considering the use of medications to assist in achieving recommended LDL-C goals [9].

The distribution of the fat allowance has been altered to recognize the value of monounsaturated and polyunsaturated fatty acids. By replacing saturated fats (cheese, whole milk, red meat) with monounsaturated fats (olive, canola oil) and polyunsaturated fats (corn oil, peanuts), LDL-C is reduced. Although replacing saturated fats with a high-carbohydrate diet results in lower LDL-C, it has the adverse effect of raising TGs and lowering HDL-C. Saturated and trans-unsaturated fatty acids should be avoided [9].

The ATP III suggests the addition of plant stanols (hydrogenated phytosterols) to the patient's diet when initial attempts to alter the diet have not resulted in reaching the LDL-C goal. Plant stanols interfere with small intestine absorption of intestinal and biliary cholesterol. While they lower LDL-C, they have no significant effect on HDL-C or TGs [10]. Phytosterols can be found in many products, including margarine spreads. Other sources of phytosterols include sesame seeds and peanuts; soybeans are a natural source of phytosterols.

Physical inactivity is an independent risk factor, raising the risk of a cardiovascular event twofold [11]. Aerobic exercise raises HDL-C and lowers TGs. When it results in weight loss, it contributes to LDL-C reduction. Weight loss also improves insulin sensitivity and serum glucose uptake, reducing the risk of DM. Cigarette smoking remains a cardiovascular risk factor. Patients who stop smoking can expect an increase of up to 30 percent in their HDL-C levels [12].

Lipid Lowering Agents

Statins

The HMGCoA reductase inhibitors, or statins, are the most widely used and arguably the most effective lipid-lowering agents currently available. These drugs work well as monotherapy and in combination with other agents. All statins lower LDL-C by upregulating hepatic LDL-C receptors after transient competitive inhibition of the reductase enzyme responsible for conversion of HMG-CoA to mevalonic acid during cholesterol synthesis. By upregulating LDL-C receptors, statins not only increase LDL-C clearance from the plasma, but also reduce circulating levels of VLDL-C and intermediate-density lipoprotein cholesterol that serve as substrates for LDL-C synthesis. When used as monotherapy, statins reduce LDL-C by 24%-60% and TGs by 22%-45%. Statins can also raise HDL-C up to 12%.

All statins reduce TC in a dose-related manner. Research has shown that doubling the dose of a statin above the minimal effective dose leads to an approximately 6% greater reduction in LDL-C [13]. Within the class of statins, differences in absorption and metabolism can be observed for each drug. Lovastatin absorption is increased significantly in the presence of food while pravastatin is best taken on an empty stomach. Because rates of cholesterol synthesis increase between midnight and 3:00 AM, patients should be advised to take lovastatin with the evening meal, while pravastatin should be taken at bedtime. The other statins are unaffected by food and can be taken at any time during the day. The statins also differ in their metabolic pathways. Of the 5 currently available statins, 3 are metabolized through the cytochrome P450 3A4 enzyme system—atorvastatin, simvastatin, and lovastatin— while fluvastatin is metabolized through cytochrome 2C9 [14]. Pravastatin is not metabolized via cytochrome P450; it undergoes hepatic sulfation [14]. Metabolism via the P450 system can result in drug interactions as other agents compete or interfere with P450 enzymatic activity.

Statins are well tolerated by most people. The most common adverse effects of the statins include gastrointestinal upset, abdominal pain, flatulence, constipation, myalgia (defined as muscle ache or weakness without elevations in creatine kinase), flu-like symptoms, and headache. Less common problems are an asymptomatic elevation of liver enzymes (defined as >3 times the upper limit of normal) and myositis (defined as muscle pain, soreness, or weakness with serum creatine kinase concentrations of >10 times the upper limit of normal), rash, insomnia, unpleasant or vivid dreams, and difficulty sleeping or concentrating [14].

Elevated hepatic transaminases generally occur in 0.5%-2.0% of cases and are dose-dependent [15]. Whcther transaminase elevation with statin therapy constitutes true hepatotoxicity has not been determined and progression to liver failure specifically due to statins is exceedingly rare. Reversal of transaminase elevation is frequently noted with a reduction in dose, and elevations rarely return when challenged with the same or different statin [16]. Furthermore, no direct evidence suggests that statins exacerbate liver disease [17]. The incidence of elevated liver enzymes can be increased in patients receiving drugs metabolized by the cytochrome P450 system. In patients exhibiting elevated liver enzymes, aminotransferase levels usually are elevated 3 times the upper limit of normal [14]. An LDL-C that suddenly declines despite stable dosing is also a sign of toxicity regardless of aminotransferase levels [17].

The ability of statins to produce myopathy (defined as any disease of the muscles) under some circumstances is well established. A common complaint is myalgia or nonspecific muscle aches or joint pains that are generally not associated with significant increases in creatine kinase. In placebo-controlled trials, the incidence of these complaints is generally similar to that reported with placebo. Myositis is even less common, occurring in 0.8%-0.9% of patients taking statins as monotherapy [17-20]. Myositis is most likely to occur in people with multiple morbidities and/or those taking several medications.

The Pleiotropic Effects of Statins

The progression of stenoses is reduced by statin treatment in CHD and it would be reasonable to assume that this direct effect on the anatomical atherosclerotic lesion is responsible for the reduction in the cardiovascular end-points. However, these effects of statins on the coronary anatomy are less impressive than the observed clinical effects. Furthermore, the onset of the clinical benefits seen with statins occurs early after the start of the statin treatment and before an effect on coronary stenosis can be demonstrated. Accordingly it has been suggested that statins have an effect on CHD that is not related to their cholesterol-lowering effect and this is referred to as the pleiotropic effect of statins.

Endothelial dysfunction is responsible for the instability of the atherosclerotic plaque, leading to plaque rupture and the acute coronary syndromes. Statins have effects on the vascular endothelium, smooth muscle, haemostatic factors and the vascular wall, which counteract the adverse effects of cardiovascular risk factors [21, 23]. It is likely that these pleiotropic effects play a substantial role in the favourable cardiovascular profile attributed to statins. These effects differ between statins and it is not unreasonable to assume that statins may not be equally effective in the prevention of vascular disease.

Nitric oxide (NO) is an important modulator of endothelial function, whereby normal arteries vasodilate in response to augmented demand and ischaemia. Endothelial function is impaired in the presence of hypercholesterolaemia and other risk factors such as smoking and the presence of the insulin resistance seen in T2D. Numerous studies have demonstrated an improved endothelial function after the use of statins. Statins increase NO via numerous mechanisms including an increased activity and induction of endothelial nitric oxide synthase (eNOS), the prevention of the hypoxia-induced down-regulation of eNOS, and the reversal of the inhibitory effect of oxidised LDL-C on NO [21, 24]. Mechanical stress stimulates smooth muscle cells (SMCs) to proliferate and to manufacture macromolecules that contribute to the extracellular matrix of the vessel wall and also to the atherosclerotic plaque. Cellular proliferation is also a process intrinsic to the atherosclerotic plaque. The lipophilic statins, but not the hydrophilic statin pravastatin, inhibit SMCs proliferation and migration induced by growth factors. The ultimate effect of this inhibition remains to be defined and indeed, the permissive effect of pravastatin on SMCs growth may be beneficial in terms of stability of the plaque's fibrous cap.

Rupture of the vulnerable atherosclerotic plaque results in the formation of a thrombus, involving platelets and various clotting factors, including tissue factor. Tissue factor expression by cultured human macrophages is suppressed by statins [25]. Platelets of patients with dyslipidaemia are more sensitive to aggregating factors and this is normalised by statins, possibly due to an alteration of the cholesterol content of platelet membranes [26].

Fibrinogen, an important predictor of CHD, is variously affected by statins, with the majority of them lowering the plasma levels [27]. However, several researchers have found that atorvastatin increases fibrinogen levels [28]. Plasminogen activator inhibitor type-1 (PAI-1), another important marker of cardiac events, is also variously affected by statins [29]. Due to a similarity in structure, lipoprotein-a (Lp(a)) competes with fibrinogen for binding on fibrinogen receptors, fibrinogen and fibrin and it is presumed that this effect is responsible for the atherogenicity of Lp(a). Lp(a) levels are increased by statins by as much as 35% [30]. However, after the reduction of LDL-C by statins, this increase in Lp(a) may be of no consequence.

Atherosclerosis has been established as a chronic inflammatory disease involving T-cells and macrophages [22]. These cells secrete proteinases and inflammatory cytokines that contribute to the instability and rupture of the vulnerable plaque. Statins contribute to the stability of the atherosclerotic plaque via reduction of the cholesterol content of the core of the plaque, by reducing the number of T-cells and macrophages, and by increasing the collagen content of the fibrous cap [29]. C-reactive protein has been identified as being an important marker and predictor of the vulnerable atherosclerotic plaque and acute coronary syndromes. Numerous studies including 4S and WOSCOPS have shown that statins are able to reduce C-reactive protein levels. Abundant data is also available to indicate that statins have an inhibitory effect on the inflammatory cells of the vulnerable plaque and inhibit the release of inflammatory cytokines [22, 23].

Statins are also able to reduce the susceptibility of LDL-C to oxidisation, a crucial step in the uptake of LDL-C by macrophages to form foam cells [22]. The use of statins seems to have an antihypertensive effect that may be clinically significant [31]. Statins have been shown to stimulate bone formation in experimental animals [32]. This has been supported by epidemiological data that links statin use with a decreased risk of fracture. However, other large studies have not been able to show any benefit on bone health with the use of statins [33]. Indeed, it has been shown that the effect of statins on bone varies with the dosage used, that statins increase osteoclast activity and bone resorption and they reduce bone mineral density in rodents [34]. The implications of these findings in humans remain to be seen.

Niacin

Niacin (nicotinic acid) is a safe, effective, low-cost, lipid-lowering therapy that has been used to treat dyslipidemia since 1955. Niacin inhibits the mobilization of free fatty acids from peripheral tissues, reducing hepatic synthesis of TGs and secretion of VLDL-C and inhibiting conversion of VLDL-C to LDL-C. The expected maximum reduction in LDL-C and VLDL-C concentrations is 30% and 40%, respectively. Niacin also increases serum HDL-C by up to 30%, surpassing all other drugs in this regard.

The most common adverse reactions associated with niacin result from prostaglandin-mediated vasodilation and include flushing, tingling, itching, rash, and headaches. Although these adverse events cannot be completely avoided, they can be minimized by slowly titrating the daily dose upward or with pretreatment using aspirin or nonsteroidal anti-inflammatory drugs. Gastrointestinal adverse reactions, such as dyspepsia, diarrhea, flatulence, and nausea, are also seen with niacin. At high doses of sustained-release niacin (>2 g daily), elevations in hepatic enzymes and uric acid can occur, although this phenomenon is not observed with high doses (up to 6 g daily) of the immediate-release formulation of niacin.

Impaired glucose tolerance has been associated with niacin therapy. However, treatment with niacin has been shown to be an appropriate treatment option for patients with diabetes. A recently published 16-week, placebo-controlled study randomized patients with diabetes into 2 groups: one receiving 1000 or 1500 mg daily of extended-release niacin and the other, placebo. In addition to noting dose-dependent increases in HDL-C and reductions in TGs, hemoglobin HbA1c levels were unchanged by niacin therapy compared to placebo. The authors concluded that low-dose extended-release niacin therapy was a safe and effective treatment for patients with T2D [35].

The extended-release form of niacin does not appear to cause the hepatotoxicity associated with previous sustained-release forms of the drug. Niacin can be dosed once, or, more commonly, twice daily. However, flushing can still be a bothersome side effect with this product. At the maximum recommended dose of 2 g daily, extended-release niacin reduces LDL-C by 16.7%, TGs levels by 34.5%, and raises HDL-C by 25.8%.

A combination of niacin with laropiprant, a flushing pathway inhibitor, has recently been added to the therapeutic choices of dyslipidemia [36]. Nicotinic acid/laropiprant is indicated for the treatment of dyslipidaemia, particularly in patients with combined mixed dyslipidaemia (characterised by elevated levels of LDL-C and TGs and low HDL-C) and in patients with primary hypercholesterolaemia (heterozygous familial and non-familial). Clinical studies have shown nicotinic acid/laropiprant to reduce LDL-C, raise HDL-C, and decrease TGs. In clinical studies, patients taking nicotinic acid/laropiprant experienced significantly less moderate-to-extreme flushing than with nicotinic acid. In a pool of four active- or placebo-controlled clinical trials of more than 4,700 patients the percentage of patients taking nicotinic acid/laropiprant who discontinued due to any flushing related symptom was 7.2 percent compared to 16.6 percent for the pooled nicotinic acid (prolonged release formulation) alone groups [36].

Fibrates

Fibrates (fibric acid derivatives) gemfibrozil and fenofibrate possess minimal LDL-C reducing capacity, but are especially effective in patients who have severe hypertriglyceridemia and low HDL-C. Fibrates are also useful in patients with combined forms of hyperlipidemia. Although the mechanism by which fibrates reduce TGs is poorly understood, it is believed these agents work to increase fatty acid oxidation in the liver resulting in decreased secretion of VLDL-C and increased LPL activity in skeletal muscle. Thus, the lipid-modifying effects of fibrates are secondary to changes in these LDL-C precursors. In patients with very high TGs, fibrates may increase LDL-C.

Fibrates may be used in combination with niacins or bile acid sequestrants as these drugs appear to be additive in lowering LDL-C and TGs and raising HDL-C. When fibrates are given in combination with a bile acid sequestrant, the administration of the 2 drugs must be separated by at least 2 hours to ensure full bioavailability of the fibrates.

Combinations of fibrates with statins are very effective in lowering LDL-C and increasing HDL-C, particularly in patients with mixed hyperlipidemia characterized by elevated TGs and LDL-C. Combinations of statins and fibrates have been historically limited by the increased risk for myopathy. This is particularly true for the combination of a statin and gemfibrozil. This risk appears to be minimized by substituting one of the newer fibrates, fenofibrate, for gemfibrozil. When the choice is made to use a statin and fibrates in combination, these drugs should be used only in the lowest effective doses and only used in patients who have normal liver and kidney function.

Bile Acid Sequestrants

Bile acid sequestrants, or resins, are now used primarily as adjuncts to statins or niacin. The resins bind to bile acids in the intestine, impeding their reabsorption.

The interruption of the entero-hepatic bile circulation causes the liver to convert more of the hepatic cholesterol pool into bile acids. It also stimulates cholesterol synthesis, resulting in increased VLDL-C production. Reduction in hepatic cholesterol pools results in an up-regulation of hepatic LDL-C receptors and increased LDL-C clearance from the circulation.

Bile acid sequestrants reduce LDL-C by 15%-27%. When used in combination with a statin, additional lowering of LDL-C of 9%-16% can be seen with as little as 4 g twice daily. Caution should be used when treating patients who have mixed hyperlipidemias, as bile acid sequestrants can increase TGs by 7%. Currently available bile acid sequestrants include cholestyramine and colestipol, which are both supplied as powdered products that require mixing with juices or food to make them more palatable.

The most common adverse reactions are gastrointestinal effects, including abdominal pain, belching, bloating, constipation, gas, heartburn, and nausea. These resins should be administered 2 hours before or 6 hours after warfarin, theophylline, digoxin, and levothyroxine. Colesevelam is a new bile acid binding resin available in tablet form. Compliance may be improved with this formulation, and no drug interactions have been noted between colesevelam and the agents known to interact with cholestyramine and colestipol, which are listed above.

Cholesterol-Absorption Inhibitors

Ezetimibe, the first agent available in this class, reduces plasma cholesterol by selectively inhibiting the absorption of dietary and biliary cholesterol from the intestine. Ezetimibe appears to specifically inhibit free cholesterol uptake into the intestinal enterocyte by interacting with a cholesterol transporter, however its exact mechanism of action remains to be elucidated [36].

Treatment of 820 patients who have hypercholesterolemia with 10 mg daily of ezetimibe reduced LDL-C by 18%, elicited a 12% decrease in TC, a 4% decrease in TGs, and a 1% increase in HDL-C. The safety profile of ezetimibe was similar to placebo, with no clinically significant changes in creatine kinase or hepatic transaminase levels and no effect on the absorption of fat-soluble vitamins A, D, and E. The most common side effects in both the active treatment and placebo groups were headache, upper respiratory tract infections, and back pain, although the back pain was not believed to be related to either ezetimibe or placebo [38].

The combination of ezetimibe and a low dose statin inhibit both the endogenous and exogenous production of cholesterol. Subjects with hypercholesterolemia treated with simvastatin 10 mg daily for 14 days achieved a 35% reduction in LDL-C levels, while those treated with simvastatin 10 mg plus ezetimibe 10 mg achieved a reduction of 52%, which is a decrease expected from the much higher simvastatin dose of 80 mg [39]. Additionally, ezetimibe has virtually no effect on statin pharmacokinetics. Thus, combining ezetimibe with a statin may reduce the dose of the latter drug required to achieve target levels, or, in patients who respond poorly to statins, may improve the lipid-lowering effect.

Diabetic Dyslipidemia

The basic feature of diabetic dyslipidemia is the elevation of TGs. Other elements of atherogenic dyslipidemia are to a large extent associated with hypertriglyceridemia. In diabetes the elevated levels of lipoproteines rich in TGs, both originating form the small intestine (chylomicrons) and endogenous (VLDL), result in an increased and prolonged postprandial hypertriglyceridemia, which in turn enhances exposition of the vessel wall to atherogenic action of these lipoproteins. The activity of both lipoprotein and hepatic lipases (LPL and HL) taking part in TG catabolism is altered in diabetes – LPL activity is partially inhibited, and HL activity up-regulated – which accelerates the IDLs to LDLs conversion, causing however no significant rise in the levels of the latter.

Insulin resistance is associated with the lack of inhibition of the VLDL synthesis. Normally their production is selectively down-regulated in postprandial state as the result of an insulin activity. This mechanism is distorted in T2D patients – insulin mediated inhibition of large VLDLs (VLDL-1) synthesis in the postprandial period becomes selectively impaired [40, 41].

Another mechanism that may take part in the pathological overproduction of VLDL particles in diabetes is the insulin triggered increased activity of sterol regulatory element-binding protein 1c (SREBP-1c). SREBP-1c activation leads to an increased *de novo* lipogenesis and plays an important role in the lipid acumulation in the liver (steatosis of the liver) thus increasing the amount of TGs available for VLDL synthesis [42].

The abundance in the plasma TGs enriched lipoproteins (TGRL) triggers the cholesterol ester transfer protein (CETP) protein dependant mechanism of lipid transfer between different classes of lipoproteins. In hypertriglyceridemia an “overflow” of TGs from TGRL enhances the CETP mediated triglyceride transport to LDL-C and HDL-C particles in exchange for cholesterol esters. As a result of HLs activity which is increased in diabetes, TGs are then split from these particles. Thus, formed LDL-C particles have a smaller cholesterol ratio per one particle of apolipoprotein B-100. They are called small dense LDL-C (subfraction LDL-B) and are believed to be more atherogenic. The atherogenic action of small dense LDL-C is dependent on the particle size, its architecture, and biological properties. Because of their small diameter, the dense LDL-C penetrate easier to a vessel wall and are more susceptible to oxidation. Small LDL-C has lower affinity to the LDL-C receptor, resulting in a prolonged plasma circulation which enhances their transport into a vessel wall and accumulation in atherosclerotic plaques. Many studies have shown that in diabetic patients the number of small dense LDL-C increases in correlation with triglyceridemia, starting from level just slightly higher than 130 mg/dL.

Low HDL

Low level of HDL-C, together with hypertriglyceridemia constitute a dominant feature of diabetic dyslipidemia. Characteristically there is a predominance of HDL-C particles of smaller diameter, belonging to the HDL

3 subclass, and large HDL-C (HDL 2) level is decreased [43]. The diameters of HDL-C correlate with the TGs in the fasting state. Another specific feature is a relatively greater decrease in the level of HDL-C containing apoAI and apoAII (Lp AI:AII). In T2D the level of Lp AI:AII seems to be more important than that of LpAI for the activity of a reverse cholesterol transport from the peripheral tissues. Therefore, it may be speculated that in T2D the low HDL-C is associated with its smaller effectiveness in the process of reverse cholesterol transport (RCT). The key mechanism responsible for HDL-C decrease is the process of lipid exchange between VLDL-C, HDL-C, and LDL-C, described above, and mediated by CETP protein. The factor intensifying this process is an increase in VLDL-C, during prolonged postprandial lipemia. TGs enriched HDL-C undergoes hydrolysis caused by HL. As a result of the interaction with liver lipase the HDL-C particles become like those of LDL-C, smaller and denser. Small dense HDL-C is catabolized faster leading to renal excretion of apoprotein AI [44, 45]. The relative predominance of the subfraction of small dense HDL-C in diabetes is another feature of diabetic dyslipidemia.

Prolonged Postprandial Lipemia

An important element of abnormal lipid metabolism in diabetes is prolonged hyperlipemia. A number of mechanisms have been suggested to be responsible for this process. The postprandial increase in concentration of chylomicrons in serum is combined with the increased levels of VLDL-C particles, the production of which, in contrast to normal state, is not inhibited in diabetes. Both types of lipoproteins compete for binding with LPL what prolongs their catabolism. Additionally, LPL activity itself is impaired. The third postulated mechanism is due to the impaired liver clearance of remnants resulting from a disturbed uptake of these particles by LDL-C receptors and LDL-C receptor related protein [46, 47]. It has not been determined whether in T2D the production of apo B-48-containing lipoproteins (chylomicrons) is enhanced. The equal atherogenic potential of remnants derived from VLDL-C (containing protein apo B-100) and remnants containing apo B-48 is emphasized. The increased postprandial lipemia is a state associated with an impaired endothelial function [48, 49]. Many studies have shown, that in the state of insulin resistance, there is the “lipid intolerance”, presenting as substantially higher increase in TGs in the postprandial period as well as long lasting fasting hypertriglyceridemia. Another important pathogenic mechanism associated with the postprandial hiperlipemiais an increased prothrombotic activity, which can be the trigger factor for an acute coronary syndrome, following fatty meal [50].

Diabetic patients show specific disturbaces in their lipid profile – known as diabetic dyslipidemia. It should be emphasized that usually in DM the level of cholesterol contained in LDL-C – widely believed to be the main atherogenic fraction of cholesterol in serum – is only slightly increased. This relatively "normal" level of LDL-C was, until recently, the reason for neglecting hypolipemic treatment of patients with diabetes. Apart form a moderate increase in LDL-C, usually not exceeding 130–150 mg/dL, in atherogenic dyslipidemia there is an increase in TGs to above 150–200 mg mg/dL and simultaneous decrease in HDL-C to below 40 mg/dL. An important additional feature of atherogenic diabetic dyslipidemia, not evaluated in standard lipidogram, is an increase of subfraction of small dense LDL-C. Recently, several simple laboratory tests have been developed to enable a quantitative assessment of LDL-B (small dense LDL-C) subfraction in daily clinical practice [51]. In hypertriglyceridemia assessment of non HDL-C may be useful. The ATP III NCEP has decided the non-HDL-C to be the secondary target of intervention in the population of patients with hypertriglyceridemia > 200 mg/dL, especially in patients with diabetes and/or the metabolic syndrome. The level of non HDL-C reflects better the risk associated with the presence of the increased levels of lipoproteins containing atherogenic apoprotein B (apo B) i.e. VLDL-C, IDL-C and LDL-C, and Lp(a). In diabetic patients the non-HDL-C target should be < 130 mg/dL.

Effects of Statins in T2D Patients: Clinical End-Point Trials

The Diabetes Atorvastatin Lipid Intervention study (DALI) compared the effects of aggressive and standard lipid-lowering therapy on fasting TGs in 217 patients with T2D [52] who were randomly assigned to aggressive (80 mg/d) or standard (10 mg/d) therapy with atorvastatin *vs.* placebo. During the 30-week study, fasting TGs were reduced by 35% with aggressive therapy and by 25% with moderate therapy compared with an increase of 10% with placebo. Results of A Randomized, Double-Blind Study to Compare Rosuvastatin and Atorvastatin in Patients with Type II Diabetes study (ANDROMEDA), in which rosuvastatin (10 or 20 mg/d) was administered to

450 patients with T2D and dyslipidemia, found that LDL-C was reduced by 51% and 57%, respectively, and 94% and 96% of patients achieved their LDL-C goal [53].

Numerous primary and secondary prevention trials provide strong evidence that statins decrease the risk of cardiovascular events in patients with DM [54-61]. These cardiovascular benefits are closely linked to the lipid-altering effects of the statin and support recommendations by the NCEP ATP III [9] for aggressive use of statins as first-line therapy in the treatment of diabetic dyslipidemia.

Subanalyses of four secondary prevention trials [Scan dinavian Simvastatin Survival Study (4S; two analyses), Cholesterol and Recurrent Events (CARE), Long-Term Intervention with Pravastatin in Ischemic Disease (LIPID), and the Heart Protection Study (HPS)] incorporating more than 9000 patients with mostly T2D or impaired fasting glucose indicate that statin treatment significantly reduces the coronary event rate *vs.* placebo in this population, and that the risk reductions are similar to those in statin-treated nondiabetic patients [54-56]. The relative risk reduction in these studies ranged from 13–55% compared with placebo. In all of these studies, statin therapy was well tolerated.

Among four primary prevention studies [Antihypertensive and Lipid-Lowering Treatment to Prevent Heart Attack Trial (ALLHAT-LLT), Anglo-Scandinavian Cardiac Outcomes Trial–Lipid Lowering Arm (ASCOT–LLA), CARDS, and Die Deutsche Diabetes Dialyze (4D)] including more than 9000 patients, only CARDS reported an unequivocal significant reduction in cardiovascular event rates [57-60]. The lack of treatment effect in ASCOT LLA is probably attributable to an inadequate number of absolute events in the smaller subgroup with diabetes [57], because the 16% reduction in myocardial infarction was not significantly different from that observed in the overall ASCOT-LLA population. The negative outcome in the Antihypertensive and Lipid-Lowering Treatment to Prevent Heart Attack Trial study is probably a result of a high rate of LDL-C-lowering treatments in the usual care group, resulting in a small (11%) difference in LDL-C concentrations between the treated and usual care groups [58]. Finally, the Die Deutsche Diabetes Dialyze trial enrolled patients with end-stage renal disease, and thus the lack of significant improvement in CHD risk rates suggests either that statin treatment may need to be provided to patients with T2D at an earlier stage of disease or that cardiovascular events in patients with end-stage renal disease is not amenable to statin therapy [59]. The results of CARDS, in which a 40% reduction in LDL-C was associated with a 37% reduction in the primary end point of time to first acute CHD event, revascularization, or stroke, does provide strong support for the view that primary prevention is beneficial in patients with T2D [60].

CHOOSING THE OPTIMAL REGIMEN

A controversy in the lipid community is whether to increase the dose of a statin or add adjunctive therapy for further LDL-C reduction. Proponents of increasing the statin dose argue that keeping the regimen simple will improve adherence, be more cost-effective, and that adjunctive agents may not provide additional pleiotropic effects comparable to higher statin doses. Conversely, others argue that doubling the statin dose may result in only a 6% further reduction in LDL-C with increased side effect potential, whereas the addition of ezetimibe or a bile acid sequestrant may result in approximately a 20% reduction in LDL-C.

However, while increasing the statin dose may be the simplest option in certain cases, statins do have dose-dependent side effects particularly when titrated to the highest doses. In cases such as this, adding a second agent (ie, ezetimibe, colesevelam) with a different site of action will not only provide more LDL-C reduction but also limit potential side effects. The use of fixed combination lipid-altering products (ie, ezetimibe/simvastatin) offers potential advantages and may be preferential to adding a separate second agent or titrating the statin. The attributes of these products compared with statin monotherapy include an overall improved effect on the lipid profile and the possibility of greater cost-effectiveness.

There is less controversy surrounding additional agents for other types of dyslipidemia. Among patients with low HDL-C, attaining the LDL-C goal is the first priority followed by achieving the non-HDL-C goal and maximizing therapeutic lifestyle changes. If HDL-C still remains a concern, therapy with niacin or fibrates may then be considered.

If patients have mixed dyslipidemia, and TGs exceed 500 mg/dL, the first objective is to reduce the TGs in order to prevent pancreatitis. Many practitioners prefer fibrates for hypertriglyceridemia because of the greater effectiveness, lower incidence of side effects, and lesser need for titration compared with niacin. These individuals may require a statin for LDL-C reduction after the TGs are reduced. Additional precautions must be taken with this combination to avoid possible adverse events.

REFERENCES

[1] American Heart Association. 2005. Heart disease and stroke statistics – 2005 update. Dallas, Texas, US: AHA.

[2] Expert Panel on Detection, Evaluation and Treatment of High Blood Cholesterol in Adults - Adult Treatment Panel III (ATP-III). Third report of the National Cholesterol Education Program (NCEP) expert panel on detection, evaluation, and treatment of high blood cholesterol in adults. *Circulation*, 2002 106, 3143–3421.

[3] Talbert RL. Hyperlipidemia. In: DiPiro JT, Talbert RL, Yee GC, Matzko GR, G. Wells B, Posey LM, editors. *Pharmacotherapy: A Pathophysiologic Approach*. 6th ed. New York: McGraw-Hill; Medical Publishing Division; 2005.

[4] Dietschy JM. Dietary fatty acids and the regulation of plasma low density lipoprotein cholesterol concentrations. *J. Nutr.* 1998 128, 444S–448S.

[5] Mahley RW; Bersot TP. Chapter 35: Drug therapy for hypercholesterolemia and dyslipidemia. In: Brunton LL, editors. *Goodman and Gilman's The Pharmacological Basis of Therapeutics*. 11th edition. New York: McGraw-Hill Companies, Inc.; 2006.

[6] Malloy MJ, Kane JP. Agents used in hyperlipidemia. In: Katzung BG. *Basic and Clinical Pharmacology*. 9th ed. New York: The McGraw-Hill Companies, Inc.; 2004:561–575.

[7] Nofer JR; Kehrel B; Fobker M; Levkau B; Assmann G; von Eckardstein A. HDL and arteriosclerosis: beyond reverse cholesterol transport. *Atherosclerosis,* 2002 161, 1–16.

[8] Downs JR; Clearfield M; Weis S; Whitney E; Shapiro DR; Beere PA; Langendorfer A; Stein EA; Kruyer W; Gotto AM Jr. Primary prevention of acute coronary events with lovastatin in men and women with average cholesterol levels: results of AFCAPS/TexCAPS. Air Force/Texas Coronary Atherosclerosis Prevention Study. *JAMA* 1998 279, 1615–1622.

[9] Expert Panel on Detection, Evaluation, and Treatment of High Blood Cholesterol in Adults. Executive summary of the third report of the National Cholesterol Education Program (NCEP) Expert Panel on Detection, Evaluation, and Treatment of High Blood Cholesterol in Adults (Adult Treatment Panel III). *JAMA,* 2001 285, 2486-2497.

[10] Miettinen TA; Puska P; Gylling H; Vanhanen H; Vartiainen E. Reduction of serum cholesterol with sitostanol-ester margarine in a

mildly hypercholesterolemic population. *N. Engl. J. Med.,* 1995 333, 1308-1312.

[11] Berlin JA; Colditz GA. A meta-analysis of physical activity in the prevention of coronary heart disease. *Am J Epidemiol* 1990 132, 612-628.

[12] Criqui MH; Wallace RB; Heiss G; Mishkel M; Schonfeld G; Jones GT. Cigarette smoking and plasma high-density lipoprotein cholesterol. The Lipid Research Clinics Program Prevalence Study. *Circulation,* 1980 62, (4 pt 2):IV70-6.

[13] Davidson M; Ma P; Stein EA; Gotto AM Jr; Raza A; Chitra R; Hutchinson H. Comparison of effects on low-density lipoprotein cholesterol and highdensity lipoprotein cholesterol with rosuvastatin versus atorvastatin in patients with type IIa or IIb hypercholesterolemia. *Am J Cardiol,* 2002 89, 268-275.

[14] Knopp RH. Drug treatment of lipid disorders. *N. Engl. J. Med.,* 1999 341, 498-511.

[15] Hsu I; Spinler SA; Johnson NE. Comparative evaluation of the safety and efficacy of HMG-CoA reductase inhibitor monotherapy in the treatment of primary hypercholesterolemia. *Ann. Pharmacother.,* 1995 29, 743-759.

[16] Cressman MD; Hooquewerf BJ; Moodie DS; Loin JW; Weinstein CE. HMG-CoA reductase inhibitors. A new approach to the management of hypercholesterolemia. *Clev. Clin. J. Med., 1988* 55, 93-100.

[17] Pasternak RC; Smith SC Jr; Bairey-Merz CN; Grundy SM; Cleeman JI; Lenfant C; American College of Cardiology; American Heart Association; National Heart, Lung and Blood Institute. ACC/AHA/NHLBI clinical advisory on the use and safety of statins. *Circulation,* 2002 106, 1024-1028.

[18] Bradford RH; Shear CL; Chremos AN; Dujovne C; Downton M; Franklin FA; Gould AL; Hesney M; Higgins J; Hurley DP, et al. Expanded Clinical Evaluation of Lovastatin (EXCEL) study results. I. Efficacy in modifying plasma lipoproteins and adverse event profile in 8245 patients with moderate hypercholesterolemia. *Arch. Intern. Med.,* 1991 151, 43-49.

[19] Illingworth DR; Tobert JA. A review of clinical trials comparing HMG-CoA reductase inhibitors. *Clin. Ther.,* 1994 16, 366-385.

[20] Pierce LR; Wysowski DK; Gross TP. Myopathy and rhabdomyolysis associated with lovastatin-gemfibrozil combination therapy. *JAMA,* 1990 264, 14-17.

[21] Maron D; Fazios LM. Current perspectives on statins. *Circulation,* 2000 101, 207–213.

[22] Ross R. Mechanisms of disease – atherosclerosis – an inflammatory disease. *N. Engl. J. Med.,* 1999 340, 115–126.

[23] Kolovou G. The Treatment of Coronary Heart Disease: An Update:Part 3: Statins Beyond Cholesterol Lowering. *Curr. Med. Res. Opin.,* 2001 17, 34–37.

[24] Sumi D; Hayashi T; Thakur NK; Jayachandran M; Asai Y; Kano H; Matsui H; Iguchi A. A HMG-CoA reductase inhibitor possesses a potent anti-atherosclerotic effect other than serum lipid lowering effects – the relevance of endothelial nitric oxide synthase and superoxide anion scavenging action. *Atherosclerosis,* 2001 155, 347–357.

[25] Lacoste L; Lam JY; Hung J; Letchacovski G; Solymoss CB; Waters D. Hyperlipidemia and coronary disease. Correction of the increased thrombogenic potential with cholesterol reduction. *Circulation,* 1995 92, 3172–3177.

[26] Soma MR; Donetti E; Parolini C; Mazzini G; Ferrari C; Fumagalli R; Paoletti R. HMG CoA reductase inhibitors. *In vivo* effects on carotid intimal thickening in normocholesterolemic rabbits. *Arterioscler. Thromb.,* 1993 13, 571–578.

[27] Behar S, for the BIP study Group. Lowering fibrinogen levels; clinical update. *Blood Coagul Fibrinolysis,* 1999 10(Suppl), S41–S43.

[28] Marais AD; Firth JC; Bateman ME; Byrnes P; Martens C; Mountney J. Atorvastatin: an effective lipid-modifying agent in familial hypercholesterolemia. *Arterioscler. Thromb. Vasc. Biol.,* 1997 17, 1527–1531.

[29] Rosenson RS; Tangney CC. Antiatherothrombotic properties of statins: implications for cardiovascular event reduction. *JAMA,* 1998 279, 1643–1650.

[30] Haffner S; Orchard T; Stein E; Schmidt D; LaBelle P. Effect of simvastatin on Lp(a) concentrations. *Clin. Cardiol.,* 1995 18, 261–267.

[31] Borghi C; Prandin MG; Costa FV; Bacchelli S; Degli Eposti D; Ambroisioni E. Use of statins and blood pressure control in treated hypertensive patients with hypercholesterolemia. *J. Cardiovasc. Pharmacol.,* 2000 35, 549–555.

[32] Mundy G; Garrett R; Harris S; Chan J; Chen D; Rossini G; Boyce B; Zhao M; Gutierrez G. Stimulation of bone formation in vitro and in rodents by statins. *Science,* 1999 286, 1946–1949.

[33] Mundy GR. Statins and their potential for osteoporosis. *Bone,* 2001 29, 495–497.

[34] Maritz FJ; Conradie MM; Hulley P; Hough FS. Statins increase quantitative histomorphometric parameters of bone formation and resorption, and decrease bone density in rodents. *Arterio. Thromb. Vasc. Biol.,* 2001 21, 1636–1641.

[35] Grundy SM; Vega GL; McGovern ME; Tulloch BR; Kendall DM; Fitz-Patrick D; Ganda OP; Rosenson RS; Buse JB; Robertson DD; Sheehan JP; Diabetes Multicenter Research Group. Diabetes Multicenter Research Group. Efficacy, safety, and tolerability of once-daily niacin for the treatment of dyslipidemia associated with type-2 diabetes: results of the assessment of diabetes control and evaluation of the efficacy of niaspan trial. *Arch. Intern. Med., 2002* 162, 1568-1576.

[36] Olsson AG; Laropiprant plus niacin for dyslipidemia and prevention of cardiovascular disease. *Expert Opin. Pharmacother*, 2010 11,1715-1726.

[37] van Heek M; Farley C; Compton DS; Hoos L; Alton KB; Sybertz EJ; Davis HR Jr. Comparison of the activity and disposition of the novel cholesterol absorption inhibitor, SCH58235 and its glucuronide, SCH60663. *Br. J. Pharmacol.,* 2000 129, 1748-1754.

[38] Knopp RH; Gitter H; Truitt T, et al. Ezetimibe reduces low-density lipoprotein cholesterol: results of a phase III randomized, double blind, placebo-controlled trial [abstract]. *Atherosclerosis,* 2001 2:38.

[39] Kosoglou T; Meyer I; Musiol B, et al. Pharmacodynamic interaction between the new selective cholesterol absorption inhibitor SCH 58235 and simvastatin [abstract]. *Atherosclerosis,* 2000, 151, 135.

[40] Brown AM; Gibbons GF. Insulin inhibits the maturation phase of VLDL assembly *via* a phosphoinositide 3-kinase-mediated event. *Arterioscler. Thromb. Vasc. Biol.*, 2001, 21, 1656–1661.

[41] Malmström R; Packard CJ; Caslake M; Bedford D; Stewart P; Yki-Järvinen H; Shepherd J; Taskinen MR. Defective regulation of triglyceride metabolism by insulin in theliver in NIDDM. *Diabetologia*, 1997, 40, 454–462.

[42] Lewis GF; Carpentier A; Adeli K; Giacca A. Disordered fat storage and mobilization in the pathogenesis of insulin resistance and type 2 diabetes. *Endocr. Rev.*, 2002 23, 201–229.

[43] Syvanne M; Ahola M; Lahdenpera S; Kahri J; Kuusi T; Virtanen KS; Taskinen MR. High density lipoprotein subfractions in non-insulin-dependent diabetes mellitus and coronary artery disease. *J. Lipid. Res.*, 1995 36, 573–582.

[44] Golay A; Zech L; Shi MZ; Chiou A; Reaven GM; Chen YD. High density lipoprotein (HDL) metabolism in noninsulin-dependent diabetes mellitus: measurement of HDL turnover using tritiated HDL. *J. Clin. Endocrinol. Metab.*, 1987 65, 512–518.

[45] Pietzsch J; Juselius U; Nitzsche S; Hanefeld M. *In vivo* evidence for increased apolipoprotein A-I catabolism in subjects with impaired glucose tolerance. *Diabetes*, 1998 47, 1928–1934.

[46] Goldberg IJ; Kako Y; Lutz EP. Responses to eating: lipoproteins, lipolytic products and atherosclerosis. *Curr. Opin. Lipidol.*, 2000 11, 235–241.

[47] Tomkin GH; Owens D. Abnormalities in apo B-containing lipoproteins in diabetes and atherosclerosis. *Diabetes Metab. Res. Rev.*, 2001 17, 27–43.

[48] Ceriello A; Taboga C; Tonutti L; Quagliaro L; Piconi L; Bais B; Da Ros R; Motz E. Evidence for an independent and cumulative effect of postprandial hypertriglycerydemia and hyperglycemia on endothelial dysfunction and oxidative stress generation. *Circulation*, 2002 106, 1211–1218.

[49] Rongen GA; Tack CJ. Triglycerides and endothelial function in type 2 diabetes. riglycerides and endothelialfunction in type 2 diabetes. *Eur. J. Clin. Invest.*, 2001 31, 560–562.

[50] Anderson RA; Jones CHJ; Goodfellow J. Is the fatty meal a trigger for for acute coronary syndromes. *Atherosclerosis*, 2001 159, 9–15.

[51] Hirano T; Ito Y; Koba S; Toyoda M; Ikejiri A; Saegusa H; Yamazaki J; Yoshino G. Clinical significance of small dense low-density lipoprotein cholesterol levels determined by the simple precipitation method. *Arterioscler. Thromb. Vasc. Biol.*, 2004 24, 558–563.

[52] Diabetes Atorvastin Lipid Intervention (DALI) Study Group. The effect of aggressive versus standard lipid lowering by atorvastatin on diabetic dyslipidemia: the DALI study: a double-blind, randomized, placebo-controlled trial in patients with type 2 diabetes and diabetic dyslipidemia. *Diabetes Care*, 2001 24 1335–1341.

[53] Betteridge D; Gibson M. Effect of rosuvastatin and atorvastatin on LDL-C and CRP levels in patients with type 2 diabetes: results of the ANDROMEDA study. *Atherosclerosis*, 2004 5, 107–108.

[54] Haffner SM; Alexander CM; Cook TJ; Boccuzzi SJ; Musliner TA; Pedersen TR; Kjekshus J; Pyo¨ra¨la¨ K. Reduced coronary events in simvastatin treated patients with coronary heart disease and diabetes or

impaired fasting glucose levels: subgroup analyses in the Scandinavian Simvastatin Survival Study. *Arch. Intern. Med.*, 159, 1999 2661–2667.

[55] Keech A; Colquhoun D; Best J; Kirby A; Simes RJ; Hunt D; Hague W; Beller E; Arulchelvam M; Baker J; Tonkin A; LIPID Study Group. Secondary prevention of cardiovascular events with long-term pravastatin in patients with diabetes or impaired fasting glucose: results from the LIPID trial. *Diabetes Care*, 2003 26, 2713–2721.

[56] Heart Protection Study Collaborative Group. MRC/BHF Heart Protection Study of cholesterol-lowering with simvastatin in 5963 people with diabetes: a randomised placebo-controlled trial. *Lancet,* 2003 361, 2005–2016.

[57] Sever PS; Dahlöf B; Poulter NR; Wedel H; Beevers G; Caulfield M; Collins R; Kjeldsen SE; Kristinsson A; McInnes GT; Mehlsen J; Nieminen M; O'Brien E; Ostergren J; ASCOT investigators. Prevention of coronary and stroke events with atorvastatin in hypertensive patients who have average or lower-than-average cholesterol concentrations, in the Anglo-Scandinavian Cardiac Outcomes Trial – Lipid Lowering Arm (ASCOT-LLA): a multicentre randomized controlled trial. *Lancet,* 2003 361, 1149–1158.

[58] ALLHAT Officers and Coordinators for the ALLHAT Collaborative Research Group. Major outcomes in moderately hypercholesterolemic, hypertensive patients randomized to pravastatin vs usual care: the Antihypertensive and Lipid-Lowering Treatment to Prevent Heart Attack Trial (ALLHAT-LLT). *JAMA*, 2002 288, 2998–3007.

[59] Wanner C; Krane V; Ma¨rz W; Olschewski M; Mann JFE; Ruf G; Ritz E; German Diabetes and Dialysis Study Investigators. Atorvastatin in patients with type 2 diabetes mellitus undergoing hemodialysis. *N. Engl. J. Med.,* 2005 353, 238–248.

[60] Colhoun HM; Betteridge DJ; Durrington PN; Hitman GA; Neil HA; Livingstone SJ; Thomason MJ; Mackness MI; Charlton-Menys V; Fuller JH; CARDS investigators. Primary prevention of cardiovascular disease with atorvastatin in type 2 diabetes in the Collaborative Atorvastatin Diabetes Study (CARDS): multicentre randomised placebo-controlled trial. *Lancet*, 2004 364, 685–696.

In: Dyslipidemia ISBN: 978-1-62100-917-7
Editors: Z. Ačimovič et.al, pp. 115-146

Chapter 4

FISH PROTEINS FOR CORONARY ARTERY DISEASE

***Rakesh Sharma*[1] *and Jose Katz*[2]**
[1]Florida State University, Tallahassee, Florida 32301, USA
[2]Dr Katz Cardiology Center, NY and Columbia University, New York 10032, USA

ABSTRACT

Recent recommendations of federal and regulatory agencies suggest revisit the claims of fish oils rich in omega-3 fatty acids and benefits in prevention of cardiovascular disease. We propose the importance of fish proteins in prevention of CVD to offer better opportunity of lowering blood pressure and coronary artery disease burden. Fish proteins are analyzed by protein electrophoresis after extraction from fish. The nutrition value of fish protein components are estimated and compared with recommended daily allowances (RDA). However, benefits of fish proteins remain less known due to poor knowledge of clinical outcomes, endpoint calibrations and their food value. The current status of fish dietary research and possibility of their use in coronary artery disease as cardioprotective food is presented if the fish protein intake may be better approach over omega fatty acids as reliable, relevant dietary recommendations for CVD in future.

Keywords: Fruits, vegetables, legumes, diet, coronary disease, trials, prognosis, mortality

1. INTRODUCTION

High protein rich fish eating in diet is routine since ages of human origin [Brown 1990]. Since nineteenth century, very low cardiovascular risk of heart disease among Eksimos in Greenland, tribals in Australia and Japan remained an unsolved puzzle. In last half centaury cholesterol was identified as responsible of heart disease and lipoproteins as main culprit of heart attacks and atherosclerosis [Cote et al. 2004, Kristensen et al. 2001, Singh et al. 2011, O'Dea 1988, 1991]. Presently, omega-3 fatty acids in fish oil are choice in cardiovascular disease prevention while statins show serious side effects. Several omega fatty acid CVD prevention trials in Asia, Europe and West have documented mixed view on benefits of omega fatty therapy. Supplementation of fish protein has emerging as better source than other protein sources in cardioprotection (Bergeron et al. 1989, 1992). American heart Association (AHA), NECP ATP III has evidenced significantly the prospects of lipid lowering by dietary means. Marine *fish protein* can also benefit the heart [NCEP Report 2001]. Recent studies provide little or no direct evidence about how fish proteins and amino acids taurine, methionine can prolong life or prevent or delay CAD events (Yamori et al. 2009). Today, it is open question. Author proposes 'one amino acid-one fatty acid concept' that amino acid (in fish protein) may be alternative approach to keep high HDL or low (LDL) or (LDL+TG+C) lipids.

In light of these recommendations, we explore the possibility of fish proteins as emerging option of cardioprotective food while most of diets and foods remain at large unidentified without scientific reason of their action. Fish proteins in food practice and major mechanistic action in control of lipids is quite unknown [Singh et al 2010]. Present paper describes the methods and results of proteins in commonly used fish in central India. Fish are excellent source of protein. 100 g cooked fish provides 18-20 g of protein (or one third RDA) containing abundant essential amino acids. The protein content in most of the fish are constant, invariable with season. Author reported the variation of fish proteins based on fish feed and month of breeding in different seasons [Shriniwas et al. 2008].

The present chapter describes introduction to fish food as source of proteins beneficial in cardiac protection, protein and amino acid composition of fish foods, new one amino acid-one fatty acid concept (one amino acid + one fatty acid can lower lipid + stabilize heart), intervention trials and futuristic approach.

2. Fish Proteins and Amino Acids

The essential amino acid compositions of fish are given in Tables 1-3 indicate abundance of lysine, methionine, threonine and low collagen. It enhances the digestibility coefficient near 100 and fish intake 100 g /day supplied around 25% RDA of adult male. However, less known fact is that fish protein consumption how decreases the atherogenic risk of vascular diseases. Very less known is the results of one study showing sole fish diet increased blood HDL lipoprotein relative to other soy proteins or milk (Beauchesne-Rondeau et al. 2010). Nonprotein nitrogen compounds are common in fish mostly in sarcoplasm which includes free amino acids, peptides, amines, amine oxides, guanidine compounds, quaternary ammonium molecules, nucleotides and urea (Ackman, 1995). Creatine is main component in fish protein. Creatine plays an important role in fish muscle in metabolism in its phosphorylated from. Endogenous proteases yield some free amino acids such as Glycine and taurine, histidine (more than 1%) in red muscle of tuna fish. Important fact of proper cooking of mahi-mahi, mackerel, skipjack and tuna fish here contributes as improper cooking only converts histidine to histamine. Histamine is not affected by heat and becomes hazard to consumers due to "scombroid poisoning"(Arino et al. 2003,Food and Drug Administration, 1989, Haard, 1995).

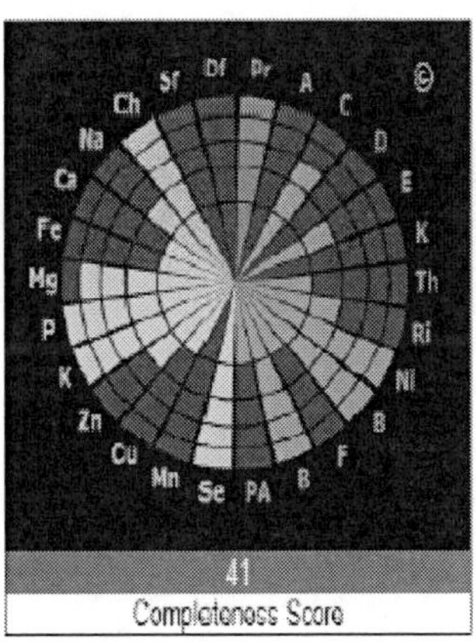

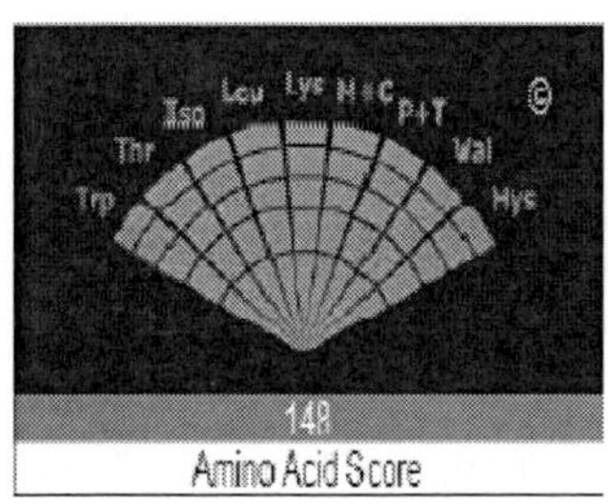

Figure 1. Nutrition value of fish protein is shown as micronutrients amino acid composition. Notice the scores of nutrition value 41 for micronutrients and 148 for amino acids.

Table 1. Apparent crude protein and amino acid availabilities (%) from various fish in cardioprotection to meet FAO/WHO daily recommended values*

Amino Acid	Atlantic salmon[a]	Coho salmon[b]	Cherry salmon[c]	Channel catfish[d]	Rainbow trout[a]	Herring menhaden
Alanine	6.52	6.08	6.35	6.31	6.57	
Arginine	6.61	5.99	6.23	6.67	6.41	9.13
Aspartate	9.92	9.96	9.93	9.74	9.94	8.54
Cystein	0.80	0.95	1.23	1.34	0.86	0.80
Glutamine	14.31	15.25	15.39	14.39	14.22	
Glycine	7.41	7.31	7.62	8.14	7.76	
Histidine	3.02	2.99	2.39	2.17	2.96	
Isoleucine	4.41	3.70	3.96	4.29	4.34	8.93
Leucine	7.72	7.49	7.54	7.40	7.59	9.05
Lysine	9.28	8.64	8.81	8.51	8.49	8.41
Methionine	1.83	3.53	3.14	2.92	2.88	
Phenylalanine	4.36	4.14	4.63	4.14	4.38	8.92
Proline	4.64	4.76	4.33	6.02	4.89	
Serine	4.61	4.67	4.48	4.89	4.66	
Threonine	4.95	5.11	4.63	4.41	4.76	8.51
Tryptophan	0.93	1.40	0.83	0.78	0.93	7.27
Tyrosine	3.50	3.44	3.58	3.28	3.38	
Valine	5.09	4.32	4.85	5.15	5.09	8.97

* FAO/WHO recommends for 80 kg man/60 kg woman are: Isoleu 800/600; Leu 1120/840; Lys 960/720; Meth/Cys 1040/780; Phenylala/Tyr 1120/840; Thr 560/420; Tryp 280/210; Val 800/600.

Table 2. Energy and fat contents in different fish

Fish (84 g/3 oz)	Calories	Total fat* (g/%DV)	Saturated fat* (g/%DV)	Protein (g)
Cod	100	½	0/0	20
Halibut	110	2/3	0/0	23
Mackerel	210	13/20	1.5/8	21
Salmon[1]	160	7/11	1/5	22
Salmon[2]	130	4/6	1/5	22
Salmon[3]	180	9/14	1.5/8	23
Salmon[4]	200	12	3	21
Tuna	320	24/30	1.2/6	21

Source: US FDA data source for daily 2000 calorie diet.

Serving size: 3 Oz skinless cooked fish portion.

* Nutrient value / % Daily Value of nutrient.

Table 3. Nutritive Profiles of fish (in 100 gm whole flesh) in cardioprotection

Fish	Calories	protein	Fat	Ca	Iron	Vit A	Vit D	EPA	DHA
Salmon/Sake	167	20.7	8.4	14	0.9	200	1300	492	820
Tuna/maguro	322	21.4	24.6	11	1	100	720	1290	2880
Tuna red meat	133	28.3	1.4	5.0	2.0	20	210	27	115
Sardine/Iwashi	213	19.2	13.8	70.0	1.7	60	390	1390	1140
Bonito/Katsuo	129	25.8	2.0	10.0	1.9	17.0	400	78	310
Mackerel/Saba	230	19.8	16.5	22.0	1.5	100	440	1210	1780
Cod/tara	70.0	15.7	0.4	42.0	0.6	100	--	37	72

Table 4. Other potential fish/ non-fish foods available in India with their protein nutritive values in cardiac prevention

Fish foods	Energy (kcal)	Protein (g)	Carbohydrates (g)	Fat (g)	Ca (mg)	P (mg)	Fe (mg)
Bhetki (fresh)	79	14.9	3.0	0.8			
Crab (muscle)	59	8.9	3.3	1.1			
Crab (small)	169	11.2	9.1	9.8			
Hilsa	273	21.8	2.9	19.4	180	280	2.1
Katla	111	19.5	2.9	2.4	530	235	0.9
Mackerel	93	19.9	0.5	2.7	429	305	4.5
Pomfrets Black	111	20.3	1.5	2.6	286	306	2.3
Pomfrets White	87	17.0	1.8	1.3	200	290	0.9
Surumai	92	19.9	NIL	1.4	92	161	2.0
Lobster	90	20.5	NIL	0.9			
Prawn	89	19.1	0.8	1.0			
Rohu	97	16.6	4.4	1.4			
Sardine	101	21.0	NIL	1.9	90	360	2.5
Shrimp(small, dried)	349	68.1	NIL	8.5			

Source: NNMB Report, 2007: National Institute of Nutrition(ICMR).

Salmon fish is oily fish, high in protein, omega-3 fatty acids and vitamin D. It is abundant in Atlantic, Pacific oceans, farm raised roe and raw sashimi in origin. Salmon fish contains dioxanes, polychlorinated biphenyl compounds. Salmon flesh is red in color but natural color is white due to astaxanthin or canthaxantin carotinoid antioxidant pigments. (Food Standards Agency, 2004, Mozaffarian 2006).

Sardines are an excellent low calorie, low carbohydrate fish food. Small, silver colored fish is also known as pilchard as a healthy snack or regular meal due to the various nutrients of fish. Pilchards are oily fish, very rich in omega-3 fatty acids, no carbohydrates, rich sources of fish protein, vitamins, A, B, C, D and E with various minerals such as iron, calcium, selenium, magnesium, phosphorous and potassium. Sardine fish contains the antioxidant, coenzyme Q10 and supports Heart Health. Oily sardine fish contains omega-3 fatty acids play a vital role in regulation of blood cholesterol levels to promote heart health. Fish intake helps to lower bad cholesterol and maintain good cholesterol levels. The improved ratio between good and bad cholesterol helps to sustain good heart health. It also reduces the risk of stroke and cardiovascular disease. Vitamin B12, also found in high concentrations in the fish, helps to protect the walls of the arteries. It promotes cardiovascular health[1]. *Tuna* is an excellent source of protein, and while some vitamin and mineral losses occur during canned tuna processing, the protein nutritive values are not dramatically changed. Tuna is an excellent source of the omega-3 fatty acids EPA and DHA, protein, potassium, selenium, and vitamin B12. Every 100 g fish food contains: Carbohydrates 0.00 g Dietary fiber 0.0 g Fat 0.95 g Protein 23.38 g Vitamin A 60 IU Thiamine (Vit. B_1) 0.434 mg Riboflavin (Vit. B_2) 0.047 mg Niacin (Vit. B_3) 9.800 mg Pantothenic acid (B_5) 0.750 mg Vitamin B_6 0.900 mg Folate (Vit. B_9) 2 mcg Vitamin B_{12} 0.52 mcg Vitamin C 1.0 mg Vitamin E 0.50 mg Vitamin K 0.1 mcg Calcium 16 mg Iron 0.73 mg Magnesium 50 mg Phosphorus 191 mg Potassium 444 mg Sodium 37 mg Zinc 0.52 mg Manganese 0.015 mg.

Cod, hake, flounder, sole contain less than 100 Kcal per 100 g fish flesh while mackerel, herring, salmon contain 250 kcal in 100 g flesh. Lean type fish have 40-60 mg cholesterol per 100 g muscle has Salmon, tuna, mackerel, sardine fish have plenty of omega 3 fatty acids also which is considered to be beneficial in cardiac prevention. Major beneficial effects of fish eating are long life expectancy, antiarrhythmic, antithrombosis, lowering triglycerides and VLDL cholesterol and antihypertensive action.

3. Fish Proteins in Cardiac Prevention

Fat intake is not culprit at all. We analyze this issue based on recent studies. The Harvard School of Public Health showed that total fat intake bore

[1] http://www.buyomegaprotein.com/benefits.aspx

no significant relation to the risk of CAD but fish intake in diet reduces the risk. Other studies suggested high omega fatty acid intake with adverse effect or stimulating fibrillation and reducing platelet aggregation or antithrombotic (Saravanan et al. 2010; Aarsetov et al. 2008, Harris 2007, Kowey et al. 2010, Pratt et al. 2010, Din et al. 2004). Investigators clearly advocated that claims of fish omega fatty acids in cardioprotection are fishy and not conclusive (Albert et al. 2010). In fact, several studies over 4 decades suggested that fish diet is rich in proteins with potentials to reduce the risk of CAD (Bernstein et al. 2010, de Leo et al. 2009, Mozaffarian 2008, Das 2000, ADA report 2003a 2003b). Saturated fats in other animal meats and foods appear to carry more risk than fish, refined carbohydrates, other food classes representing the bulk of recommended daily calories[2]. Fish diet is rich in unsaturated fatty acids (mono- and polyunsaturated fatty acids), good protein quality (45%-80%). In addition, the fish foods contain mercurial, arsenic etc contaminants and appear to have adverse effects. Several other epidemiological studies (Gordon et al. 1981) have shown no evidence that man who eat less fat live longer or have fewer myocardial infarctions (MI). Epidemiological evidence that fish fed population on low carbohydrate diet for 6 months had low fat and low risk of heart disease (Westman et al. 2002). Other group reported that fish protein was better than other cardioprotective dietary and omega-3 options (Bergeron et al. 1989, 1992). The relationship of fish muscle protein characteristics in different seasons including casein content and W-6/W-3 contents was evidenced with relationship of improved blood pressure and episodes of heart complications after fish eating. Our hypothesis was: can fish proteins are cardiac protecting? Recently, study reported protein content of H. fossilis (BLOCH) fish to evaluate the nutritive value of fish that has impact on fish eating community in Nagpur. In Western world and Southeast Asia, fish eating concept is emerging as affluent style of cardiac protection. We project that fish protein content analysis can highlight further potentials of fish eating in cardiac protection due to its good contents on omega fatty acids, fish oils and protein values in fish. Authors reported a short cross sectional study on patients with cardiac heart disease (arrhythmia, early myocardial infraction sign, high blood pressure) having treatment of betablocker and diuretics for last known 3-6 months. Patients were monitored and intervened with W-3/W-6 EPA/DHA rich oils or with fish eating for at least 100 Grams/100 Kg weight/day in last 1-3 Months. Cross sectional routine BP, Heart indicators, dietary intake of fish proteins and oil content was calculated. To further calculation of fish protein content,

[2] http://www.13.waisays.com/protein.htm

electrophoretic mobility indices were screened in the H.fossilis. The casein content, oil content and heart indicators were correlated to establish the effect of fish protein and oil content as effective cardiac protection measures. Intervention of fish eating and withdrawal of chemotherapy improved the serum lipid profile (HDL-3), BP, EKG. After 3 months fish rich diet intervention showed further improved BP, lipid profile (HDL-3), EKG finding towards normal. However, during 3-6 months intervention effect was slower. The H.fossilis protein content analysis suggested several proteins on electrophoresis bands with possibility of other protein candidates in cardioprotection. Casein protein was distinct candidate. During seasonal protein patterns, male fish protein suggested April-June and female Dec-Jan period as high protein yield cardioprotective diet.

Cardioprotection by fish eating is a dilemma despite of known fish infection, environmental hazards. Fish breeding industries have significant role to serve infection–free, protein rich fish food through out the year by measures of spawning in different seasons specially resting phase of fish. Liver oil and proteins as combined have proven as best known cardiac protection non-chemotherapeutic measures. Our growing understanding of fish eating intervention with slow withdrawal of betablockers/diuretics has become routine clinical practice. In this report, major points are: 1.W-3/W-6 Fatty acid rich oils with fish casein protein projects the enhanced cardiac protection perhaps due to casein sparing effect and reduced free radicals enough to the level of further progress of cardiac injury; 2. fish breeding with spawning in rest phase shows impact on fish protein value; 3. cardiac protection is multifactorial and enhanced by combined fish liver oil omega fatty acids and amino acid composition in fish proteins. Fish high quality proteins with fish liver oils serve best cardiac protection.

Fish meat or muscle has good protein quality. Salmon, halibut, tuna are major fish foods. Casein, Taurine are major fish proteins. In following sections, we focus on fish protein contents and their role in hypertension and heart disease.

Essential hypertension is associated with higher lipoperoxidation and imbalanced antioxidant status (means oxidative stress) with increased RBC enzymes (Yuan et al.2002). Mechanisms are not well established yet but some experimental studies have established the modulation of HDL and reverse cholesterol transport, lipoprotein lipase activity in livers and action of actinopectin, ACE inhibitory peptides heart (Hersberger et al. 2005, O'Shea et al. 2010, De Lao et al. 2009, Jung et al. 2008, Siddique et al.2004). Blocking of AHR2 and ARNT1 gene expression was reported as cardioprotective (Antkiewicz et al. 2006).

3.1. One Amino Acid-one Fatty Acid Concept

One amino acid in fish protein may be alternative approach to keep high HDL, low (LDL+TG+C) lipids for cardioprotection. Author reported the benefits of casein protein in hypercholesterolemic hamsters. In gall stone bearing hamsters with high plasma cholesterol, casein (250 g/Kg wt) showed cholesterol lowering after 4 weeks. Cholesterol lowering effect against cholesterol saturation and cholelithiasis in hamsters was shown by using cytochrome P450 and cholesterol 7α Hydroxylase biomarker proteins. In casein fed animals, plasma lipids including total cholesterol, phospholipids, lipoproteins and biliary lipids were brought down in time fashion manner (Sharma et al. 2008). To prove the cardioprotective effect of casein, following studies suggest evidences in favor:

1) Researchers from Algeria compared fish protein consumption to casein in an animal model of hypertension. The fish protein diet lowered blood pressure and plasma total cholesterol compared with casein, leading researchers to conclude fish protein attenuated the development of hypertension(Yahia et al. 2003a). Casein intake changed the tissue antioxidant status in rats (Yahia et al. 2003b).
2) Dietary cod fish consumption rich in casein showed effect on muscle physiology by insulin induced activation of glutamate transfer protein and translocation characteristics in fish muscle (Tremblay et al. 2003)
3) Bergeron et al.1989 reported the influence of fish proteins and compared to casein and soy protein for their effect on serum and liver lipids, and serum lipoprotein cholesterol levels in the rabbits. Investigators established the mechanism of cardioprotection by fish protein at molecular level as modulation of HDL and lipoprotein lipase activityCasein effect was closer with fish protein in cardioprotection and regulation of plasma lipids (Bergeron et al. 1992a, Jacques et al. 1995). Furthermore, interaction between fish proteins and lipids was responsible in regulation of serum and lipids in liver that gives benefit of cardioprotection (Bergeron et al. 1991). Fish protein actually showed effect in proportion of its incremental amounts in diet or factorial manner (Bergeron, 1992b). In following

section, individual amino acids are discussed their value in cardiac prevention.

Taurine in Fish Proteins

Taurine (T) was first noted as beneficial for stroke and cardiovascular diseases (CVD) prevention in genetic rat models, stroke-prone spontaneously hypertensive rats (SHRSP) (Yamori et al. 1984). The preventive mechanisms of T were ascribed to sympathetic modulation for reducing blood pressure (BP) and anti-inflammatory action. Recent epidemiological surveys revealed that taurine was effective in experimental arterio-lipidosis prone SHRSP selectively bred rats with higher reactive hypercholesterolemia. In fact, sulphur containing amino acids taurine, methionine showed attenuated effect on severe hypertension and stroke due to activation of 7α hydroxylase to accelerate cholesterol excretion into bile acids (Jerlich et al. 2000). Rats quickly developed arterial fat deposition and fatty liver which was attenuated by dietary Taurine supplementation (Yokogoshi et al. 1999). In fact, investigators suggested the role of rate limiting enzyme cholesterol 7α hydroxylase to bile synthesis to decrease cholesterol and inverse correlation between mRNA levels and enzyme gene CYP7A1 as contributor to increased serum triglycerides.

Methionine, Cystein, Cystine in Fish Protein

Sulphur containing amino acids are abundant in fish muscle. Their role in lowering lipid is not established but these amino acids play role in sarcoplasm in muscle cells and contractility (Ackman et al. 1995, Arino et al. 2003). Amino acid requirements in different age groups are specific as amino acid contents of fish foods stated below. In general amino acid requirement pattern is clear; methionine and cystine typically are most scarce in daily food. National Institute of Nutrition (ICMR) monitored the following requirements and suggested fish foods shown in Table 4 and Table 5 (NNMB report, 2007, FAO/WHO report).

For adults, methionine and cystine always are most scarce (100%). Even if only minimal amounts of amino acids are consumed, in 81% of foods methionine and cystine again are most scarce, in 19% tryptophan is scarce.

For schoolchildren, again methionine and cystine are most scarce in all foods except grains. In 47% of grains methionine and cystine are most scarce, but in 53% lysine is, depending on the applied amino acid requirements. However, nobody eats grains like bread and cornflakes *only*, again making methionine and cystine the scarcest amino acids in diets comprising lots of grains.

For younger children protein requirements match food-protein contents better, levelling relative availability; in 57% methionine and cystine are most scarce. In 25% phenylalanine and tyrosine are, in 13% tryptophan is, and in 5% of foods isoleucine is most scarce, depending on the applied amino acid requirements. But, again, methionine and cystine are the scarcest amino acids.

Table 5. Requirement of total amino acids available from fish foods in different age groups. Requirements as % daily values (%DV) are shown red for adults, yellow for schoolchildren and *green* in adolescents fed on fish food

herring	389	47	269	167	149	389	231	231	100
cod	482	56	256	198	167	396	233	228	100
saith	488	45	295	186	162	417	271	243	100
mackerel	398	62	278	193	147	414	251	223	100
tuna	491	67	316	233	216	482	269	262	100
trout	487	58	301	222	164	429	258	260	100
salmon	408	53	281	164	145	358	234	224	100
Brown shrimp	412	43	202	180	133	402	204	173	100
mussel	363	56	284	191	191	353	219	214	100

Guanidine, Histidine, Creatine, Glycine

Creatine is considered as essential amino acid to keep creatine kinase enzyme CK-MB with other isozymes working in cardiac muscle contraction and it is also recommended for athletic activity (Food and Drug Administration, 1989, Haard, 1995). The mechanism of guanidine ME10092 is not known in cardiac prevention but its analogue CARIPORIDE is emerging to provide cardiac prevention by sodium-hydrogen exchange inhibition in mitochondria (Oliver et al. 2004, Mentzer et al. 2008). Histidine is converted in histamine in cooked fish but gives constipation problems (Arino et al. 2003,Food and Drug Administration, 1989).

Glycine is basically a novel anti-inflammatory, immunomodulatory cytoprotective agent and believed beneficial in cardiac prevention (Haard, 1995).

Squalene in Fish

Squalene is bioactive isoprenoid substance present in shark liver oil. It is explored to neutralize the harmful effects of free radicals and improve membrane stabilizing properties. Sqalene has been proven as safe dietary supplement without any side effect (Farvin et al. 2006, Thankappan et al. 2003).

3.2. Interventional Studies

Fish proteins showed similarity in casein for short-term metabolic effects of dietary interventions on various risk factors of CAD. Casein was given as supplements in form of beta casein A1 and A2 and effect was observed on plasma lipoproteins. Major effect was seen on HDL and triglycerides (Chin-Dusting et al. 2006, Veen et al. 2006, Nilausen et al. 1999). Studies provide little or no direct evidence about how fish proteins and amino acids taurine, methionine can prolong life or prevent or delay CAD events (Yamori et al. 2009). These studies also suffer from a major design flaw that only recently has been accepted in cardiovascular medicine: Metabolic or biochemical measures can provide mechanistic insights and are essential building blocks in therapeutic development, but they cannot reliably predict the effect of proposed interventions on clinical events. Authors describe metabolic and biochemical insight on fish protein composition and muscle physiological influence on heart physiology.

*Car*diovascular *Di*seases and *A*limentary *C*omparison Study CARDIAC trial was designed on biological markers of diet with hypertension and CVD mortality (Yamashita et al. 2006). CARDIAC (CVD and Alimentary Comparison) WHO-coordinated multi-center epidemiological survey on diets and CVD risks in 61 populations showed that twenty-four-hour urinary (24 hr) Taurine excretion was inversely related significantly with coronary heart disease mortality. Higher 24U-T excreters had significantly lower body mass index, systolic and diastolic BP, total cholesterol (T-Cho), and atherogenic index (AI: T-Cho/high density lipoprotein-cholesterol) than lower T excreters. T effects on CVD risks were intensified in individuals whose 24U-T and -magnesium (M) excretions were higher. Furthermore, higher Na excreters with higher heart rate whose BP were significantly higher than those with lower heart rate were divided into two groups by the mean of 24U-T, high and low T excreters. Since the former showed significantly lower BP than the latter, T may beneficially affect salt-sensitive BP rise. In other study on 61 populations including Guiyang, China and St. John's, Newfoundland, Canada where in which the means of both 24U-T and -M were high and low, respectively. The former and the latter had low and high CVD risks, respectively (Yamori et al. 2006). Australian Aboriginals living at the coastal area in Victoria eat T- and M-rich bush and mainly sea foods and were free from CVD 200 years ago. Such diet pattern indicated that T- and/or M-containing seafood, vegetables, fruits, nuts, milk, etc was similar to prehistoric hunters' and gatherers' food should be good for CVD prevention. Now same population had highest CVD risks in those areas. In recent study, preventive effects of T good for health and longevity were first noted experimentally and later were proven epidemiologically in humans (Yamori et al 2006, 2009).

EXPEDITION study was an important, large, drug intervention study that tested the hypothesis that targeting an ischemia/reperfusion injury mechanism can result in a reduction in the incidence of MI after coronary bypass (CABG). The results of EXPEDITION revealed (1) compelling evidence that the incidence of intraoperative myocardial necrosis in patients undergoing CABG occurs more frequently than previously appreciated; (2) a reduction in the primary endpoint of all-cause death or nonfatal MI was achieved with cariporide (a guanidine analogue); and (3) the reduction in the primary endpoint was due almost exclusively to a reduction in nonfatal MI; however, this benefit was offset by a higher incidence of CVEs and mortality in patients receiving cariporide (Mentzer et al. 2008). It needs further research on other alternative amino acid or analogue with better outcome.

3.3. Nutrient Composition of Fish and Environment

There is some knowledge available to address whether fish protein composition can affect appetite or quality of life. High quality fish meal is a complete source of amino acids. It contains high levels of lysine and methionine, essential trace elements. Due to high nutrition value of fish meal, research for alternate sources of protein is continuing such as unsaturated fatty acid rich oils, legumes, by-product meats, brewer by-products with high nutritive value or available essential amino acids. The protein quality of fish is evaluated by freshness of flesh by volatile nitrogen content, intact proteins.

Methods of assessing quality fall into three catagories: 1. Inspection; 2. Chemical analysis; 3. Biological evaluation. Chemical tests of protein quality includes for: acid corrected pepsin digestible protein (AOAC and Torry methods), protein extraction methods by electrophoresis and SDS-PAGE, multienzyme digestible proteins, total volatile basic nitrogen, lysine, sulphydryl groups and disulphide bonds. Biological tests are multienzyme and Torry pepsin solubility tests. Protein data in fish after PAGE electrophoresis is shown in Table 6.

In collaborative study, electrophoresis SDS-PAGE on Excel 15% homogeneous gels was easy and cheap silver staining method was better option than urea IEF method to identify fish species and distinguish different proteins in salmon, halibut and tuna fish food extracts(Etienne et al. 2000). Species differentiation was easy in different fish family by spot analysis as shown in Figure 2. For details of the technique, readers are referred to read original paper (Etienne et al. 2000).

Biogenic amines (cadaverine, putrescine, histamine, tyramine) are used to assess fish food quality (Chen et al. 2007) as shown in Table 7. Thiamine content was used as indicator of high quality as function of thermal processing conditions in canned salmon fish industry (Quitral et al. 2006). Temperature and food processing method used are two major factors in changing fish protein specially composition of methionine, cystein, taurine in fish food (Lipka et al. 1993).

Table 6. Protein content of different fish are shown by using methods urea immunoelectrophoresis and SDS-PAGE methods

Fish species(common name)	Protein content (mg/g flesh)	
	Urea IEF	SDS-PAGE
Oncorhynchus gorbuscha(pink salmon)	182.4	215.6
Oncorhynchus keta(chum salmon)	125.8	135.2
Oncorhynchus mykiss(rainbow trout)	170.5	153.2
Salmo trutta(sea trout)	170.3	170.3
Salmo salar(Atlantic salmon)	208.1	167.1
Hippoglossus hippoglossus(halibut)	124.8	136.9
Reinhardtius hippoglossoides(Greenland halibut)	171.2	155.6
Thunnus alalunga(albacore)	185.9	197.3
Thunnus albacores(yellow fin tuna)	148.8	173.9
Katsuwonis pelamis(skipjack tuna)	139.6	179.6

Source: Modified with permission from original reference Etienne et al. 2000.

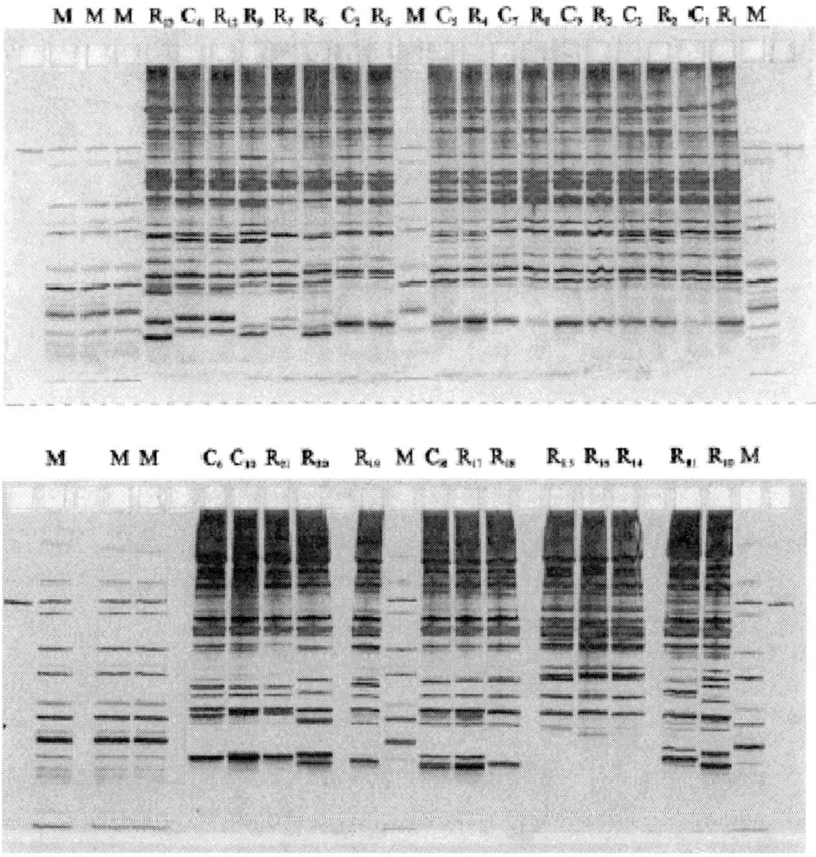

Figure 2. SDS-PAGE. Extracts of raw (references R1-R21) and cooked (samples C1-C10) fish muscle were run on Excel gel homogeneous 15%. M) p*I* calibration proteins. The cathode is at the top of the gel. Reproduced with permission from reference((Etienne et al. 2000).

Table 7. Biogenic amine content in fish meals

Fish	Histamine mg/100g[1]	Putrescine mg/100 g[1]	cadaverine mg/100 g[1]	Tyramine mg/100 g[1]
sardine	5.1	33.2	41.6	4.9
tuna	17.9	95.7	159.0	36.6

[1]expressed as a concentration of the dry matter.

Table 8. Public online resources on fish diet and proteins for cardiovascular protection

http://www.fao.org/docrep/003/w7499e/w7499e13.htm
http://linkinghub.elsevier.com/retrieve/pii/S0045653504010173
http://www.pulsus.com/CARDIOL/home.htm
www.banglajol.info/index.php/JARD/article/viewPDFInterstitial/762/800
http://linkinghub.elsevier.com/retrieve/pii/S1050464801903742
http://linkinghub.elsevier.com/retrieve/pii/S1050464807000186
http://www.blackwell-synergy.com/doi/abs/10.1111/j.1365-2761.2006.00722.x
http://ohioline.osu.edu/sc172/sc172_17.html
http://aem.asm.org/cgi/content/abstract/74/11/3551
http://findarticles.com/p/articles/mi_m0887/is_6_20/ai_75818427

Temperature around the year in different months plays a significant role in fish breeding and fish growth.

The fish protein content, amino acid composition and growth hormone action all are dependent on dietary composition of fish meal and its quality. In recent study, Shriniwas et al. 2008 reported effect on temperature on fish liver proteins identifiable in H. fossalis and C. catla as shown in Figure 3 and protein contents on PAGE-electrophoresis in catala fish (a common human fish food in Nagpur, Central India).

Investigators established better SDS-PAGE electrophoresis method (see Appendix 1).

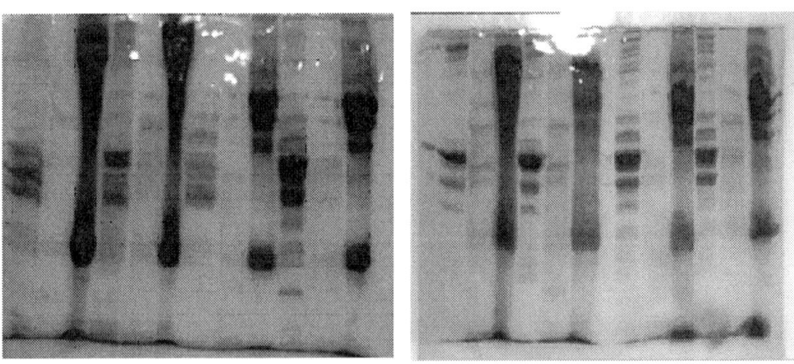

Figure 3. SDS-PAGE electrophoresis is shown for two fish species H. fossilis (A) and C.catla (B) muscle proteins.

4. Effect of Fish Dietary Proteins on Coronary Artery Disease

In a randomized, single blind , controlled intervention trial, Singh et al. 200 Indo-Mediterrarian dietary 24 weeks trial on 59 patients with high blood pressure and subsequent preliminary study Bharati D Shriniwas et al. 2008, both demonstrated effects of fish supplementation on blood pressures and blood lipids and nitric oxide in patients with hypertension and dyslipidemia. Epidemiological studies and intervention trials indicated that fish intake can decrease cardiovascular events in patients with high risk of coronary artery disease(CAD). It was evidence that total or whole fish intake can also decrease blood pressures and blood lipids. The beneficial effects of fish intake may be due to long chain w-3(EPA and DHA) fatty acids and amino acids in the fish proteins. Investigators examined the effects of mild fish intake on cardiovascular risk factors, in patients with hypertension and dyslipidemia. The effects of Indian fish (30g/day) and AHA-step 1 diet were compared for a period of 24 weeks in patients with mild hypertension(140/90-160/110mmHg) and dyslipidemia. All patients with known hypertension, with either low HDL($<$40mg/dl) or high triglycerides($>$150mg/dl) or hypercholesterolemia(T-C $>$200mg/dl), were randomly divided into fish intake group(Gr A, n=30)and control group(Gr.B, n=29) by the dietitian for computer generated data after an observation period of one week. Mean age (57.1±9.5 vs 58.5±8.7 years), body mass index (25.2±2.5 vs 24.6±2.6 kg/m^2 and waist circumferences (89.8±8.8 vs 90.8±6.8 cm) were comparable at entry to the study. Mean systolic(155.6±11.5 vs 153.7±13.4 mmHg) and diastolic (100.4±8.8 vs 101.0±7.2 mmHg) blood pressures, in fish intake group and control group, were comparable, at entry to the study. However, after 24 weeks, there was a significant fall in both mean systolic(146.3±9.5 vs 152.5±12.0 mmHg, $P<0.03$) and diastolic(92.4±9.8 vs 98.6±7.6 mmHg, $P<0.05$) blood pressures in the intervention group compared to control group, respectively. Mean concentrations of blood glucose (126.8±35.1 vs 120.5±31.8 mg/dl) were comparable at entry to the study . Plasma levels of antioxidant vitamins E (20.5 ± 3.2 vs 21.5±3.3 pmol/L),and C(19.2±3.3 vs 19.6±3.4 pmol/L) were comparable at the entry to the study . After 24 weeks, plasma levels of vitamin E (28.6±3.8 vs 21.6±3.1 pmol/L) and C (25.4±4.0 vs 19.2±3.4 pmol/L) showed significant increase and TBARS and MDA (0.88±0.26 vs 1.52 ±0.31pm/dl)showed significant($p<0.02$) decrease in the fish intake group compared to control group, respectively. HDL-C and triglycerides showed no significant differences before entry to the study. However,after 24 weeks of treatment with fish, HDL-C(40.5±7.2 vs 35.2±8.5 mg /dl ,$p<0.01$) showed a significant increase. T-C/HDL-C ratio (4.6±1.0 vs 5.6±1.1, mg/dl, $p<0.05$) and LDL cholesterol (120±12 vs 132±15,mg/dl, $P<0.04$), showed a significant decrease, while T-C and TG showed nonsignificant decline, in the intervention group compared to control group. Serum concentration of nitrite which is a indicator of nitric oxide were comparable at baseline(0.57±0.12 vs 0.63±0.16 μmol/L).After 24 weeks, there was a significant increase in nitrite levels in the intervention group compared to control group(0.93±0.21 vs 0.71±0.13 μmol/L) respectively. Increase in nitrite is an indicator of nitric oxide which is known to decrease endothelial function and blood pressures. Investigators indicated that mild fish intake can decrease blood pressures, modulate oxidative stress and increase HDL-C, T-C/HDL-C ratio and nitric oxide in patients with high risk of CAD.

4.1. Indian Fish Proteins in Lipid Lowering: One Amino Acid-One Fatty Acid Concept

Although the benefits of specific lipid- lowering drugs are unquestioned, the relation between these drug effects and the potential of diet to improve outcomes is unclear. Fish diet is emerging as cardioprotective food. Fish protein, amino acid and omega 3/6 Eicosapentanoic/Docosahexanoic fatty acids (EPA/DHA) rich fish oils are considered to reduce total blood fat LDL(Bad), cholesterol and simultaneously raise the HDL (Good) cholesterol. Eating fish (muscles) 30 gm/day reduces the chances of coronary heart disease by more than 70%. The fish proteins along with omega-3/omega-6 (EPA/DHA) polyunsaturated fatty acid content protects the heart by keeping HDL3 ratio high and rapid conversion of cholesterol to its products by Cholesterol oxidase. Shriniwas et al. 2008 analyzed diets with fish protein contents, omega 3/omega 6 (EPA/DHA) contents and essential amino acid composition and correlated diets with their lipid lowering effect. Idea was that fish muscle proteins and essential amino acid composition provide better cardiac protection along with omega 3/omega 6 fatty acids against sudden death from cardiovascular diseases. Investigators proposed possibility of 'one amino acid-one fatty acid concept' that fish proteins and amino acids may be alternative approach to keep high HDL, low (LDL+TG+C) lipids, and prevent the formation of blood clots to avoid the build up of plaque in the coronary and cardiovascular arteries specially in patients showing sensitivity to omega fatty acids. In preliminary experiment, different fish proteins, oil composition, amino acids were analyzed and compared. Borderline essential hypertensive patients with cardiac problem (High BP with risk of heart attack) were suggested to take 30 gm/day fish in diet (to keep up recommended omega 3/omega 6 (EPA/DHA) content) over six months. After follow-up, fish oil content, fish protein content, fish amino acid contents were correlated with heart indicators to establish the effect of fish protein and oil content on cardiovascular diseases. Simultaneously, the biochemical analysis of less known Indian fish protein content and protein composition were analyzed by electrophoretic mobility. In following section, we describe fish protein composition, omega fatty acids and benefits in cardiovascular prevention.

The source of Indian fish is important in fish protein quality (see Table 9). The amino acid composition and fish protein content, omega 3/omega 6 (EPA/DHA) content differ in fish (see Table 10). The progressive increased lipids lead to coronary heart disease and hyperlipidemia as main cause of death in most fast growing urban cities. Several previous studies report that fish protein, fish amino acid content were determinants to lower the blood lipids with possibility of cardiac prevention if taken in specific quantity not exceeding fat energy 10% intake. The protein content of fish *Heteropneustes fossilis* ,*Catla catla* suggested specific proteins of size 35, 41, 67 and 94 kDalton as SDS-PAGE electrophoretic bands (see Figure 3). The fish grown during December and January months possibly served as better cardiac protector fish food and served as high yield cardioprotective proteins. On an average, fish diet brings down BP by -8/2.5mmHg Systolic/Diastolic pressure. However, strong belief is in favor of fish proteins playing significant role in Okinawa community. Fish proteins and omega 3/omega 6 (EPA/DHA) appear as effective measure to have keep fat energy intake below 10% and prevent coronary heart disease and stroke by keeping minimum hypertension, B.P. and hypercholesterolemia. The main player in lipid lowering is HDL-3 molecule composed of Apolipoprotein E2 and apo-A moieties bound with major cholesterol esters. Its main component LP-A apolipoprotein has been established as cholesterol scavenger to keep low total lipid content (LDL+TG+free cholesterol). However, the link between HDL-3 synthesis and effect of fish protein intake is not established. An alternate possibility appears of specific amino acid for specific fatty acid during synthesis of HDL-3 and Lp-A protein synthesis. If it is true, it amounts to support the fish diet intake and its effect on BP and lipid lowering. The fish proteins are effective diet component in lipid lowering. The amino acid-fatty acid combination may be effective tool to synthesize HDL-3 protein-lipid composition in favor of keeping cholesterol esters rich Apo-protein moieties in HDL molecule as scavenger of cholesterol.

Different fish have different protein composition and its nutrition value. The temperature and fish feed further play important role in fish protein quality in muscle growth over months. The other additional fact can be attributed due to calcium mediated tropomyosin-actin activity in muscle contraction at low temperatures in December-January months. However much remains unknown if fish muscle and its protein composition differ during fish growth in different months of year. Also it is not established if fish eaten in different months of the year has different cardioprotection.

Table 9. Fish protein and energy contents

Indian Fish	Protein gm%	Energy Cals/gm	BP(mmHg) -Sys/Dia
Hilsa	21.8	273	-10/2.5
Rohu	16.6	97	-6/2
Catla	19.5	111	-9.5/3
Sardine	21.0	101	-5/3
Surmai	19.9	92	---
Mackerel	18.9	93	-10/2.5
Black Pomfret	20.3	111	---
White Pomfret	17.3	87	---

Table 10. Omega 3/6 fatty acid content of popular fish and EAA (%) in fish proteins

Omega 3/6 PUFA	Qty gm	Amino acid	mg/g
sardine	0.98-1.70	Lysine	8.8
Trout	0.98-0.84	Tryptophan	1.0
Anchovy	1.4	Histidine	2.0
Herring	1.2-1.8	Phenylalanine	3.9
Catfish	0.3-0.4	Leucine	8.4
Crab	0.35	Isoleucine	6.0
Lobster	0.07-0.41	Threonine	4.6
Mackerel	0.34-1.57	Methionine/cystine	4.0
		Valine	6.0

5. Effect of Fish Proteins on Atherosclerosis

Randomized, controlled trials have been conducted to know the effects of fish dietary interventions alone as a strategy to arrest the progression of CAD as shown in table 4. The lifestyle Heart Trial randomly allocated 48 patients to either a very- low- fat diet (10% of daily energy intake) and intensive lifestyle intervention (exercise, stress management, smoking cessation, group counseling) or usual care(25) .After 1 year, the experimental group showed more favorable changes in angina frequency and angiographic stenosis. After 5 years, the experimental group still had more favorable changes in low coronary stenoses and fewer cardiac events.(26)

In the St. Thomas' Atherosclerosis Regression Study (STARS), 90 men were randomized to one of three arms: usual care; a low- fat, low- cholesterol diet, high in omega- 6 and omega-3 fatty acids and high fiber (similar to AHA diet recommendations); or this diet plus cholestyramine. Patients in the fish diet- only group had significant improvements in several angiographic variables and fewer cardiac events or interventions compared with the usual-care group. Fish intake (2 meals containing fish meat weekly) may offer some promise in the secondary prevention of CAD. Fish is a functional food which is rich in n-3 fatty acids, coenzyme O10 and selenium which are protective against CAD.

6. Prevention of Coronary Artery Disease: Fish Diet

There is a strong necessity for fish dietary research to eschew the type of data that would no longer be accepted to recommend a drug for prevention or treatment of CAD. In contrast to drugs, the basic fish diet will not change in future firm biological basis, ability to reduce one or more risk factors for CAD with better clinical outcomes or epidemiological data in favor of safe and feasible accepted fish diet over a lifetime. Advanced understanding of trial methods can overturn many erroneous principles or public policy "Does a program of dietary advice that can be followed by a typical person lead to fewer cardiovascular events?" sounds effectiveness or pragmatic type to answer the public- policy question [54].

CONCLUSION

This is appropriate time to apply to fish protein diet research. We believe that indications or claims made for weight loss or health improvement via fish diet- whether made by authors, the government, or associations- must be supported by 3 types of evidence: 1.proof that the diet provides essential nutrients in actual patients; 2.efficacy studies; and 3.randomized, controlled trials with clinical events as endpoints.

APPENDIX 1. SAMPLE PREPERATION AND SDS-PAGE ELECTROPHORESIS

The muscle was homogenized in lyses buffer separately containing 50 mM tris-HCL, pH 7.5, 50mM $MgCl_2$, 1mM EDTA, 1% Triton X-100 and 1mM PMSF (Phenyl methyl sulphonyl fluoride) as protease inhibitor. The homogenized tissues were centrifuged and supernatants were taken as a protein sample. These protein samples were stored at -20^0 C and used for gel electrophoresis.

Chemicals:

1) Acrylamide stock solution (30%)

Acrylamide	29.2 Gm.
Bis Acrylamide	0.8 Gm.
Double Distilled Water (DDW)	upto 100ml.
(Stored in refrigerator in the dark)	

2) Running gel buffer (1.5 M, Tris HCl, pH 8.8) :

Tris	18.17 Gm
D.D.W.	upto 100ml

3) Stacking Gel buffer (0.05 M, Tris HCl, pH 6.8):

Tris	1.5 Gm
D.D.W.	Upto 25ml

4) 10% SDS:

SDS	10 Gm
D.D.W.	upto 100ml

5) 10% APS (Ammonium per sulphate):

APS	0.1gm
D.D.W.	upto 10ml

6) Treatment buffer (pH 6.8)

Tris	2.5ml (pH 6.8)
SDS	4.0ml
Glycerol	2.0ml
β-Mercaptoethanol	1.0ml
D.D.W.	upto 10ml
Bromophenol blue	(pinch)

7) Tank buffer

Tris	3.0 Gm
Glycine	14.4 Gm
SDS	10.0 ml
D.D.W.	upto 1000ml

8) Stain stalk (1%Coomassie brilliant blue R_{250}):

Coomassie brilliant blue	2 Gm
D.D.W.	upto 200ml

9) Working Stain:

Stain stock	12.5ml
Methanol	50ml
Glacial acetic acid (GAA)	10ml
D.D.W.	upto 100ml

10) Destaining solution (50% Methanol, 10% acetic acid):

Methanol	500ml
GAA	100ml
D.D.W.	upto 1000ml

11) Preparation of separating gel (15%):

30% Acrylamide (stock)	10ml
Tris (pH 8.8)	5.0ml
D.D.W.	4.6ml
Ammonium per sulphate	200μl
SDS	200μl
TEMED	8.0μl

12) Preparation of staking gel

30% Acrylamide (stock)	1.0ml
Stacking gel buffer (pH 6.8)	750ml
D.D.W.	4.1ml
Ammonium per sulphate	60 μl
SDS 60μl	
TEMED	6 μl

Procedure: The 15% Sodium Dodecyl Sulphate Polyacrylamide Gel Electrophoresis (SDS-PAGE) was prepared on vertical gel (16X14cm) system (Bangalore genei, India) in between two glass plates using 1mm spacer up to desired level .Stacking gel was poured in the vacant space. The comb of 1mm well was inserted in the stacking gel. After solidification of gel, the comb was taken out.

Mean while the protein sample was added with an equal volume of sample buffer (0.1M Tris-HCL, pH 6.8 ;10% glycerol; 1% sodium dodecyl sulphate, 0.02% bromophenol blue), treated with 1% β-mercaptoethanol and denatured at 98^0C for 3 min. The gel plates was attached to the electrophoretic chamber. The upper and lower tanks of chamber were filled with tank buffer.50μl of sample was taken and mixed with 50μl of treatment buffer and denatured. 30μl of sample was loaded in each well with a micropipette. A lane containing molecular weight marker (Bangalore Genei, India) was included in a well and 50mA current was supplied to electrophoretic chamber .After running the samples the gel was then stained with Coomassie Brilliant Blue $R\text{-}_{250}$ (SRL – India) and destained.

References

Ackman RG. Composition and nutritive value of fish and shellfish. In: Ruiter A(ed.) Fish and Fishery Products: *Composition, Nutritive Properties and Stability*, Willingford: CAB International. 1995, pp 117-156.

Alne H, Thomassen MS, Takle H, Terjesen BF, Grammes F, Oehme M, Refstie S, Sigholt T, Berge RK, Rørvik KA. Increased survival by feeding tetradecylthioacetic acid during a natural outbreak of heart and skeletal muscle inflammation in S0 Atlantic salmon, Salmo salar L. *J. Fish Dis.* 2009 Nov;32(11):953-61.

American Dietetic Association; Dietitians of Canada. Position of the American Dietetic Association and Dietitians of Canada: vegetarian diets. *Can. J. Diet. Pract. Res.* 2003a;64(2):62-81.

American Dietetic Association; Dietitians of Canada. Position of the American Dietetic Association and Dietitians of Canada: Vegetarian diets. *J. Am. Diet. Assoc.* 2003b;103(6):748-65.

Andersen Ø, Østbye TK, Gabestad I, Nielsen C, Bardal T, Galloway TF. Molecular characterization of a PDZ-LIM protein in Atlantic salmon

(Salmo salar): a fish ortholog of the alpha-actinin-associated LIM-protein (ALP). *J. Muscle Res. Cell Motil.* 2004;25(1):61-8.

Anderson JS. *Nutritive value of fish meal for salmonids*. pp389-393.

Andreassen R, Lunner S, Høyheim B. Characterization of full-length sequenced cDNA inserts (FLIcs) from Atlantic salmon (Salmo salar). *BMC Genomics*. 2009 Oct 30;10:502.

Antkiewicz DS, Peterson RE, Heideman W. Blocking expression of AHR2 and ARNT1 in zebrafish larvae protects against cardiac toxicity of 2,3,7,8-tetrachlorodibenzo-p-dioxin. *Toxicol. Sci.* 2006 Nov;94(1):175-82.

Arino A, Beltran JA, Roncales P. Dietary importance of fish and shellfish. In: Caballero B, Trugo L, Finglas P (eds) Encyclopedia of Food Sciences and Nutrition, 2nd edn., Oxford: Elsevier Science Ltd. 2003, pp 2471-2478.

Beauchesne-Rondeau E, Gascon A, Bergeron J, Jacques H. Plasma lipids and lipoproteins in hypercholesterolemic men fed a lipid-lowering diet containing lean beef, lean fish, or poultry. *Am. J. Clin. Nutr.* 2003 Mar;77(3):587-93.

Bergeron, N. and Jacques, H. Influence of fish protein as compared to casein and soy protein on serum and liver lipids, and serum lipoprotein cholesterol levels in the rabbit. *Atherosclerosis* 1989;78: 113–121.

Bergeron N, Deshaies Y, Jacques H. Factorial experiment to determine influence of fish protein and fish oil on serum and liver lipids in rabbits. *Nutrition.* 1992b Sep-Oct;8(5):354-8.

Bergeron N, Deshaies Y, Jacques H. Dietary fish protein modulates high density lipoprotein cholesterol and lipoprotein lipase activity in rabbits. *J. Nutr.* 1992a Aug;122(8):1731-7.

Bergeron N, Deshaies Y, Lavigne C, Jacques H. Interaction between dietary proteins and lipids in the regulation of serum and liver lipids in the rabbit. Effect of fish protein. *Lipids*. 1991 Sep;26(9):759-64.

Bernstein AM, Sun Q, Hu FB, Stampfer MJ, Manson JE, Willett WC. Major dietary protein sources and risk of coronary heart disease in women. *Circulation.* 2010 Aug 31;122(9):876-83.

Bower NI, Johnston IA. Discovery and characterization of nutritionally regulated genes associated with muscle growth in Atlantic salmon. *Physiol. Genomics.* 2010 Oct;42A(2):114-30.

Brown WV. Dietary recommendations to prevent coronary heart disease. *Ann. NY Acad. Sci.* 1990;598:376-88.

Capilla E, Díaz M, Albalat A, Navarro I, Pessin JE, Keller K, Planas JV. Functional characterization of an insulin-responsive glucose transporter

(GLUT4) from fish adipose tissue. *Am. J. Physiol. Endocrinol. Metab.* 2004 Aug;287(2):E348-57.

Chen YC, Tou JC, Jaczynski J. Amino acid, fatty acid, and mineral profiles of materials recovered from rainbow trout (Oncorhynchus mykiss) processing by-products using isoelectric solubilization/precipitation. *J. Food. Sci.* 2007 Nov;72(9):C527-35.

Chin-Dusting J, Shennan J, Jones E, Williams C, Kingwell B, Dart A. Effect of dietary supplementation with beta-casein A1 or A2 on markers of disease development in individuals at high risk of cardiovascular disease. *Br J Nutr.* 2006 Jan;95(1):136-44.

Cleveland BM, Evenhuis JP. Molecular characterization of atrogin-1/F-box protein-32 (FBXO32) and F-box protein-25 (FBXO25) in rainbow trout (Oncorhynchus mykiss): Expression across tissues in response to feed deprivation. *Comp. Biochem. Physiol. B. Biochem. Mol. Biol.* 2010 Nov;157(3):248-57.

Côté S, Dodin S, Blanchet C, Mulvad G, Pedersen HS, Blanchet C, Holub BJ, Dewailly E. Very high concentrations of n-3 fatty acids in peri- and postmenopausal Inuit women from Greenland. *Int. J. Circumpolar. Health.* 2004;63 Suppl 2:298-301.

Das UN. Beneficial effect(s) of n-3 fatty acids in cardiovascular diseases: but, why and how? *Prostaglandins Leukot Essent Fatty Acids*. 2000;63(6):351-62.

De Lao F, Panarese S, Gallerani R, Ceci LR. Angiotensin converting enzyme (ACE) inhibitory peptides: production and implementation of functional food. *Curr. Pharm. Des.* 2009;15(31):3622-43.

de Lorgeril M, Salen P, Monjaud I, Delaye J. The 'diet heart' hypothesis in secondary prevention of coronary heart disease. *Eur. Heart J.* 1997 Jan;18(1):13-8.

Etienne M, Jérôme M, Fleurence J, Rehbein H, Kündiger R, Mendes R, Costa H, Pérez-Martín R, Piñeiro-González C, Craig A, Mackie I, Malmheden Yman I, Ferm M, Martínez I, Jessen F, Smelt A, Luten J. Identification of fish species after cooking by SDS-PAGE and urea IEF: a collaborative study. *J. Agric. Food Chem.* 2000 Jul;48(7):2653-8.

Farvin KHS, Anandan R, Kumar HS, Shiny KS, Mathew S, Sankar TV, Nair PGV.Cardioprotective effect of squalene on lipid profile in isoprenaline-induced myocardial infarction in rats. *J. Med. Food* 2006; 9(4):531-536.

Fernandes J. [Nutrition and health--recommendations of the Health Council of the Netherlands regarding energy, proteins, fats and carbohydrates]. *Ned. Tijdschr. Geneeskd.* 2002 Nov 23;146(47):2226-9.

Food and Drug Administration. The Fish List, FDA Guide to Acceptable Market Names for Food Fish sold in Interstate Commerce 1988. US Government Printing Office, Washington DC; 1989,pp 30-34.

Gordon T, Kagan A, Garcia- palmieri M, et al. Diet and its relation to coronary heart disease and death in three populations. *Circulation.* 1981;63:500-515.

Haard NF. Composition and nutritive value of fish proteins and other nitrogen compounds. In: Ruiter A(ed.) Fish and Fishery Products: *Composition, Nutritive Properties and Statbility*. Wallingford: ACB International. 1995, pp 77-115.

Harris WS. Omega-3 fatty acids and cardiovascular disease: a case for omega-3 index as a new risk factor. *Pharmacol. Res.* 2007 Mar;55(3):217-23.

Heine-Bröring RC, Brouwer IA, Proença RV, van Rooij FJ, Hofman A, Oudkerk M, Witteman JC, Geleijnse JM. Intake of fish and marine n-3 fatty acids in relation to coronary calcification: the Rotterdam Study. *Am. J. Clin. Nutr.* 2010 May;91(5):1317-23.

Hersberger M, von Eckardstein A. Modulation of high-density lipoprotein cholesterol metabolism and reverse cholesterol transport. *Handb. Exp. Pharmacol.* 2005;(170):537-61.

Jacques H, Gascon A, Bergeron N, Lavigne C, Hurley C, Deshaies Y, Moorjani S, Julien P. Role of dietary fish protein in the regulation of plasma lipids. *Can. J. Cardiol.* 1995 Oct;11 Suppl G:63G-71G.

Ji H, Bradley TM, Tremblay GC. Atlantic salmon (Salmo salar) fed L-carnitine exhibit altered intermediary metabolism and reduced tissue lipid, but no change in growth rate. *J. Nutr.* 1996 Aug;126(8):1937-50.

Jordal AE, Hordvik I, Pelsers M, Bernlohr DA, Torstensen BE. FABP3 and FABP10 in Atlantic salmon(Salmo salar L.)--general effects of dietary fatty acid composition and life cycle variations. *Comp. Biochem. Physiol B. Biochem. Mol. Biol.* 2006 Oct;145(2):147-58.

Jung UJ, Torrejon C, Tighe AP, Deckelbaum RJ. n-3 Fatty acids and cardiovascular disease: mechanisms underlying beneficial effects. *Am. J. Clin. Nutr.*2008 Jun;87(6):2003S-9S.

Keli SO, Feskens EJ, Kromhout D. Fish consumption and risk of stroke. The Zutphen Study. *Stroke.* 1994 Feb;25(2):328-32.

Krauss RM, Eckel RH, Howard B, Appel LJ, Daniels SR, Deckelbaum RJ, Erdman JW Jr, Kris-Etherton P et al. AHA Dietary Guidelines: revision 2000: A statement for healthcare professionals from the Nutrition Committee of the American Heart Association. *Circulation.* 2000 Oct 31;102(18):2284-99.

Krawczenko A, Ciszak L, Malicka-Blaszkiewicz M. Carp liver DNase--isolation, further characterization and interaction with endogenous actin. *Comp. Biochem. Physiol. B. Biochem. Mol. Biol.* 2005 Jan;140(1):141-51.

Kristensen SD, Iversen AM, Schmidt EB. n-3 polyunsaturated fatty acids and coronary thrombosis. *Lipids.* 2001;36 Suppl:S79-82.

Lipka E, Ganowiak Z. Nutritional value of protein subjected to technologic processing. Changes in content of biologically active amino acids--methionine, cysteine and taurine under the influence of sterilization. *Rocz. Panstw. Zakl. Hig.* 1993;44(2-3):151-8.

Murashita K, Kurokawa T, Ebbesson LO, Stefansson SO, Rønnestad I. Characterization, tissue distribution, and regulation of agouti-related protein (AgRP), cocaine- and amphetamine-regulated transcript (CART) and neuropeptide Y (NPY) in Atlantic salmon (Salmo salar). *Gen. Comp. Endocrinol.* 2009 Jun;162(2):160-71.

Mikalsen AB, Sindre H, Torgersen J, Rimstad E. Protective effects of a DNA vaccine expressing the infectious salmon anemia virus hemagglutinin-esterase in Atlantic salmon. *Vaccine.* 2005;23(41):4895-905.

Mozaffarian D. Fish intake, contaminants and human health Evaluating the risks and the benefits. *J. Am. Med. Assoc.*2006;296(15):1885-1899.

Mozaffarian D. Fish and n-3 fatty acids for the prevention of fatal coronary heart disease and sudden cardiac death. *Am. J. Clin. Nutr.* 2008 Jun;87(6):1991S-6S.

Nalbone G, Termine E, Léonardi J, Portugal H, Lechêne P, Calaf R, Lafont R, Lafont H. Effect of dietary salmon oil feeding on rat heart lipid status. *J. Nutr.* 1988 Jul;118(7):809-17.

NCEP REPORT: Executive summary of the third report of the National Cholesterol Education Program (NCEP) Expert Panel on Detection, Evaluation, and Treatment of High Blood Cholesterol in Adults (Adult Treatment Panel III). *JAMA.* 2001;285:2486 -2497.

Nilausen K, Meinertz H. Lipoprotein(a) and dietary proteins: casein lowers lipoprotein(a) concentrations as compared with soy protein. *Am. J. Clin. Nutr.* 1999;69(3):419-25.

O'Dea K. cardiovascular disease risk factors in Australian aborigines. *Clin Exper. Pharm. Physiol.* 1991;18(2):85-88.

O'Dea K The hunter gatherer lifestyle of Australian aborigines:implications for health. In *Current Problems in Nutrition, Pharmacology and Toxicology*.Ed. McLean, Wahlqvist ML.1988 pp27-36.

Oliver DW, Dormehl IC, Wikberg JE, Louw WK, Dambrova M, et al. In vivo measurements of the cerebral perfusion and cardiovascular effects of the novel guanidine ME10092 in the non-human primate, Papio ursinus. *Life Sci.* 2004 Sep 10;75(17):2057-64.

Owen RW, Giacosa A, Hull WE, Haubner R, Würtele G, Spiegelhalder B, Bartsch H. Olive-oil consumption and health: the possible role of antioxidants. *Lancet Oncol.* 2000 Oct;1:107-12.

Quitral V, Romero N, Avila L, Marín ME, Nuñez H, Simpson R. Thiamine retention as a function of thermal processing conditions: canned salmon. *Arch. Latinoam. Nutr.* 2006 Mar;56(1):69-76.

Sales J. Prediction of digestible energy content across feed ingredients and fish species by linear regression. *Fish Physiol. Biochem.* 2009 Nov;35(4):551-65.

Santos MJ, Llopis J, Mataix FJ, Urbano G, Lopez Jurado M. Influence of dietary fish on fatty acid composition of the erythrocyte membrane in coronary heart disease patients. *Int. J. Vitam. Nutr. Res.* 1996;66(4):378-85.

Sharma R, Singh RB, Moffatt RJ, Katz J. Dietary fat intake: Promotion of disease in carotid artery disease: Lipid lowering versus side effects of statins. In: *Modern Dietary Fat Intakes in Disease Promotion*, Nutrition and Health, Ed. Meester,FD, Zibadi S, Watson, RR. Springer Science+Business Media.New York pp151-185.

Sharma R, Tandon R. Comparison of Cholesterol Lowering Diets: Apple, Casein Cytochrom P450 protein and Cholesterol 7α Hydroxylase Activities in Hamsters. Available from *Nature Precedings* http://dx. doi.org/10.1038/npre.2010.4205.1 (2010).

Shriniwas BD, Sharma R, Sharma A, Baile VV, Chen CJ, Raghukumar V(2008) Particles as Protein Markers: Nanoscale Microscopy Towards Picoscale. *Nanotechnology Research Journal.*2(3):25-32.

Siddiqui RA, Shaikh SR, Sech LA, Yount HR, Stillwell W, Zaloga GP. Omega 3-fatty acids: health benefits and cellular mechanisms of action. *Mini. Rev. Med. Chem.* 2004 Oct;4(8):859-71.

Siddiqui RA, Harvey KA, Zaloga GP. Modulation of enzyme activities by n-polyunsaturated fatty acids to support cardiovascular health. *J. Nutr. Biochem.* 2008;19(7):417-437.

Sidhu KS. Health benefits and potential risks related to consumption of fish or fish oil. *Regul. Toxicol. Pharmacol.* 2003 Dec;38(3):336-44.

Singh RB, de Meester F, Wilczynska A. The tsim tsoum approaches for prevention of cardiovascular disease. *Cardiology research and practice* 2010 doi:10.4061/2010/824938.

Sinnatamby RN, Dempson JB, Power M. A comparison of muscle- and scale-derived delta13C and delta15N across three life-history stages of Atlantic salmon, Salmo salar. *Rapid. Commun. Mass Spectrom.* 2008 Sep;22(18):2773-8.

Soriguer F, Serna S, Valverde E, Hernando J, Martín-Reyes A, Soriguer M, Pareja A, Tinahones F, Esteva I. Lipid, protein, and calorie content of different Atlantic and Mediterranean fish, shellfish, and molluscs commonly eaten in the south of Spain. *Eur. J. Epidemiol.* 1997 Jun;13(4):451-63.

Stoknes I, Rustad T. Purification and characterization of a multicatalytic proteinase from Atlantic salmon (Salmo salar) muscle. Comp Biochem *Physiol. B. Biochem. Mol. Biol.* 1995 Aug;111(4):587-96.

Thankappan TK: Isolation of squalene from shark liver oil. In: *Seafood Safety* (Surendran PK, Mathew PT, Nirmala T, Narayan N, Jose J, Boopendranath MR, Lakshmanan PT, Nair PGV, eds.), *Society of Fisheries Technologists*, Cochin, India, 2003, pp.173–175.

Tremblay F, Lavigne C, Jacques H, Marette A. Dietary cod protein restores insulin-induced activation of phosphatidylinosol 3-kinase/Akt and GLUT4 translocation to the T-tubules in skeletal muscle of high-fat-fed obese rats. *Diabetes* 2003;52:29-37.

Undeland I, Ellegard L, Sandberg AS. Fish and cardiovascular health. *Scand. J. Nutrition.* 2004; 48(3): 119-130.

van Rooyen J, Esterhuyse AJ, Engelbrecht AM, du Toit EF. Health benefits of a natural carotenoid rich oil: a proposed mechanism of protection against ischaemia/ reperfusion injury. *Asia Pac. J. Clin. Nutr.* 2008;17 Suppl 1:316-9.

Venn BJ, Skeaff CM, Brown R, Mann JI, Green TJ. A comparison of the effects of A1 and A2 beta-casein protein variants on blood cholesterol concentrations in New Zealand adults. *Atherosclerosis.* 2006 Sep;188(1):175-8.

von Schalburg KR, Yazawa R, de Boer J, Lubieniecki KP, Goh B, Straub CA, Beetz-Sargent MR, Robb A, Davidson WS, Devlin RH, Koop BF. Isolation, characterization and comparison of Atlantic and Chinook salmon growth hormone 1 and 2. *BMC Genomics.* 2008 Nov 3;9:522.

Westman EC, Yancy WS, Edman JS, et al. Effect of 6- month adherence to a very low carbohydrate diet program. *Am. J. Med.* 2002;113:30-36.

"What's an oily fish?" Food Standards Agency .2004-06-24. http://www.food.gov.uk/news/newsarchive/2004/jun/oilyfishdefinition.

Xiao X, Qin Q, Chen X. Molecular characterization of a Toll-like receptor 22 homologue in large yellow croaker (Pseudosciaena crocea) and promoter activity analysis of its 5'-flanking sequence. *Fish Shellfish Immunol.* 2011 Jan;30(1):224-33.

Yahia DA, Madanni S, Prost E, Prost J, Bouchenak M, Belleville J. Tissue antioxidant status differs in spontaneously hypertensive rats fed fish protein or casein. *J. Nutr.* 2003a;133:479-482.

Yahia DA, Madani S, Savelli JL, Prost J, Bouchenak M, Balleville J Dietary fish protein lowers the blood pressure and alters tissue polyunsaturated fatty acid composition in spontaneously hypertensive rats. *Nutrition.* 2003b;19:342-346.

Yamori Y. The stroke-prone spontaneously hypertensive rat: Contribution to risk factor analysis and prevention of hypertensive diseases. In: de Jong W (ed). *Handbook of Hypertension.* Elsevier, Amsterdam.1984:240–255.

Yamori Y, Liu L, Mizushima S, Ikeda K, Nara Y and CARDIAC Study Group. Male cardiovascular mortality and dietary markers in 25 population samples of 16 countries. *J. Hypertens* 2006;24:1499–1505.

Yamori Y Hypertension and biological dietary markers in urine and blood: A progress report from the CARDIAC study group. In: Yamori Y and Strasser T (eds). *New Horizons in Preventing Cardiovascular Diseases.* Elsevier, Amsterdam:1989 p111–126.

Yamori Y, Liu L, Mori M, Saara M, Murakami S, Nara Y, Mizushima S. Taurine as the nutritional factor for the longevity of the Japanese revealed by a world-wise epidemiological survey. *Adv. Exp. Med. Biol.* 2009; 643:13-25.

Yamashita M, Konagaya S. Purification and characterization of cathepsin L from the white muscle of chum salmon, Oncorhynchus keta. *Comp Biochem. Physiol. B.* 1990;96(2):247-52.

Yuan YV, Kitta DD, Godin DV. Heart and red blood cell antioxidant status and plasma lipids in the spontaneously hypertensive and normotensive wistar Kyoto rats.*Can J Physiol Pharmacol.* 2002;74:290-297.

Zhang J, Sasaki S, Amano K, Kesteloot H. Fish consumption and mortality from all causes, ischemic heart disease, and stroke: an ecological study. *Prev. Med.* 1999 May;28(5):520-9.

Zheng X, Tocher DR, Dickson CA, Bell JG, Teale AJ. Highly unsaturated fatty acid synthesis in vertebrates: new insights with the cloning and characterization of a delta 6 desaturase of Atlantic salmon. *Lipids.* 2005 Jan;40(1):13-24.

Zhong Z, Wheeler MD, Li X, Froh M, Schemmer P, Yin M, Bunzendaul H, Bradford B, Lemasters JJ. L-Glycine: a novel antiinflammatory, immunomodulatory, and cytoprotective agent. *Curr. Opin. Clin. Nutr. Metab. Care.* 2003 Mar;6(2):229-40.

In: Dyslipidemia ISBN: 978-1-62100-917-7
Editors: Z. Ačimovič et.al, pp. 147-174

Chapter 5

QUALITATIVE AND QUANTITATIVE CHARACTERISTICS OF LOW-DENSITY LIPOPROTEINS IN ATHEROSCLEROSIS-RELATED DISEASES

Jelena Vekic*, Aleksandra Zeljkovic, Zorana Jelic-Ivanovic, Vesna Spasojevic-Kalimanovska and Slavica Spasic

Department of Medical Biochemistry, Faculty of Pharmacy, University of Belgrade, Belgrade, Serbia

ABSTRACT

Elevated serum low-density lipoprotein cholesterol (LDL-cholesterol) concentration is firmly established as a risk factor for atherosclerosis and cardiovascular disease (CVD). Yet, a number of CVD patients have LDL-cholesterol levels within the recommended range, suggesting the need for advanced lipid testing to determine residual risk. In this manner, further improvement of risk assessment might be accomplished through a more detailed insight into qualitative and

* Corresponding author: Jelena Vekic, Department of Medical Biochemistry, Faculty of Pharmacy, University of Belgrade, Vojvode Stepe 450, P. Box 146, 11 000 Belgrade, Serbia, Tel: +381 11 39 51 284, Fax: +381 11 39 72 840, E-mail: jelena.vekic@pharmacy.bg.ac.rs

quantitative characteristics of LDL particles. Plasma LDL population comprise complex spectrum of particles of different size, density and lipid composition. There is now ample of evidence that certain LDL subclasses, particularly small, dense LDL particles, are superior to LDL-cholesterol in terms of CVD risk prediction. Furthermore, elevation of small, dense LDL particles is also observed in various atherosclerosis-related conditions, such as end-stage renal disease or ischemic stroke. Whereas the measurement of LDL-cholesterol concentrations has proven clinical utility, the usefulness of LDL particles characterisation in clinical practice needs to be further explored. Understanding of structural complexity of LDL particles and their functional consequences might improve prevention and prognosis of atherosclerosis-related diseases, leading toward more specialized therapeutic approaches. Therefore, the questions of whom, when and how to asses small, dense LDL particles still remain open.

INTRODUCTION

Prevention of atherosclerosis development is one of major challenges for contemporary medicine and science. Over the last century the knowledge of the complex molecular mechanisms involved in atherogenesis has been greatly enhanced and available evidence suggests that dyslipidemia is deeply involved in the development of atherosclerosis and its associated complications, such as cardiovascular disease (CVD).

The role of low-density lipoproteins (LDL) in atherosclerotic process has been well documented through many clinical and experimental studies. High LDL-cholesterol concentration has been recognised as a main lipid risk factor for CVD, while the maintenance of its optimal level is an imperative of modern cardiovascular medicine. However, CVD risk prediction, if judged only via LDL-cholesterol levels, may misclassify considerable number of patients with elevated risk. It is often underappreciated that measuring lipids is really a surrogate for measuring the lipoproteins that actually carry lipids in circulation and act in atherogenesis. Even if this surrogate relationship between plasma lipid and lipoproteins has been described more than forty years ago [1], contemporary laboratory monitoring of dyslipidemia has been barely changed ever since.

Heterogenic nature of plasma lipoproteins was recognized in the early 1950's. Starting from this pioneer study of Gofman and Lindgren [2] until today, the knowledge on lipoprotein diversity has been steadily accumulating. Years of investigations revealed several molecular mechanisms responsible for such remarkable heterogeneity. Also, it has been confirmed that different qualitative and quantitative characteristics of LDL subfractions considerably influence degree of atherogenicity of this lipoprotein. Besides, these studies also exposed certain risk groups in the general population. Current investigations are focused to establish whether prognostic information derived from LDL-cholesterol concentrations might be improved by additional evaluation of qualitative and quantitative characteristics of LDL subclasses.

Qualitative and Quantitative Characteristics of LDL Particles

As a general rule, LDL particles are described as lipoproteins within the density limits of 1.019-1.063 kg/L and size ranging from 20 to 30 nm. The size and density are determined by the amount and content of lipids within the particles. The hydrophobic core contains more cholesterol esters than triglycerides and is surrounded by a hydrophilic shell of free (unesterified) cholesterol and phospholipids. Structural integrity of the particles is maintained by apolipoprotein B-100 (apoB) [3]. Each LDL particle contains only one apoB molecule and is incapable of exchanging it with other lipoproteins due to the nature of lipid/protein interactions on the surface.

LDL plasma population is composed of heterogeneous particles, differing in size, density and lipid content [4], which can be separated by a variety of analytical techniques. For instance, by ultracentrifugal separation, it is possible to resolve up to seven distinct subfractions [5]. However, the most convenient approach is a classification into four major subclasses (LDL I-IV). LDL I and LDL II subclasses comprise largest, but less dense particles, whereas LDL III and LDL IV are presented by small, dense particles. If judged only by size, LDL plasma population can be divided into small (22.0-25.5 nm), intermediate (25.6-26.5 nm) and large (26.6-28.5 nm) particles [6]. From the clinical standpoint, the most important are small, dense LDL particles ($\leq$ 25.5 nm).

A remarkable LDL particles diversity arises from the changes in the amount of their precursors, as well as in the activities of the enzymes involved in lipoprotein metabolism. Generally, it appears that generation of small, dense LDL particles could be a consequence of: overproduction of very-low-density lipoproteins (VLDL) [7], reduced lipoprotein-lipase and increased hepatic-lipase activities [8], but also due to remodeling processes mediated by cholesterol-ester transfer protein (CETP) [9]. More likely, production of small, dense LDL particles involves all of previously mentioned mechanisms. It has been suggested that the VLDL-LDL delipidation cascade consists of parallel metabolic pathways responsible for the formation of different LDL subclasses. Interestingly, the largest VLDL particles have been identified as the precursors of the smallest and densest LDL subclasses [9]. The role of lipolytic enzymes in modulation of LDL particle size has been also confirmed [10]. Activity of hepatic lipase, the key enzyme involved in the hydrolysis of triglycerides in LDL particles, in cooperation with CETP significantly contributes to the increase of small, dense LDL particles. Thus, serum triglyceride concentration seems to be an important modulator of LDL subclass distribution [11], since CETP-mediated transfer of triglycerides from VLDL to LDL in hypertriglyceridemic state ultimately leads to the formation of smaller, triglyceride-rich LDL particles [8].

Austin and Krauss [12] employed mathematical model to convert densitometric profiles of LDL subclasses into Gausian curves and revealed that plasma LDL particles possess bimodal distribution. Based on this typical appearance, the authors suggested classification in two distinct phenotypes. Accordingly, predominance of large, floating LDL I and LDL II particles (>25.5 nm) was designated as LDL A phenotype, whereas the predominance of small, dense LDL III and LDL IV particles (≤ 25.5 nm) refers to LDL B phenotype [12]. The expression of LDL B phenotype is genetically determined (heritability range is 35-45%) and based on an autosomal dominant or codominant model with varying polygenic effects [13]. As a result, the prevalence of LDL B phenotype can reach up to 35% in healthy adult men, while the frequency is 5-10% in young men and 15-25% in postmenopausal women [14]. Epidemiological studies revealed numerous environmental factors that may influence the expression of this phenotype in asymptomatic population, including abdominal obesity and metabolic syndrome [14], dietary habits [15], alcohol intake [16] and use of hormone replacement therapy or oral contraceptives [17].

Plasma LDL-cholesterol concentration has proven clinical utility and is traditionally used as a surrogate measure for circulating LDL particles. However, based on all previous observations, it is obvious that LDL particles composition may vary in terms of cholesterol content. Hence, plasma LDL-cholesterol concentration cannot provide sufficient information on quality and quantity of LDL particles.

Clinical application of LDL subclasses assessment is limited by the availability and cost of laboratory techniques. LDL particles characterisation requires either special equipment or a lengthy analytical time; therefore it is still unsuitable for general clinical use. To date investigators have described several different approaches to separate LDL subclasses based on a variety of their physicochemical properties including particle density and size, charge and lipid composition (Table 1).

Table 1. The most commonly used separation techniques and nomenclatures for LDL subclasses

Gradient gel electrophoresis		Ultracentrifugation		NMR	
Name	Size (nm)	Name	Density (kg/L)	Name	Size (nm)
LDL I	27.2-28.5	LDL 1	1.019-1.023	L 3	20.0-20.4
LDL IIA	26.5-27.2	LDL 2	1.023-1.028		
LDL IIB	25.5-26.5	LDL 3	1.028-1.034		
LDL IIIA	24.7-25.5	LDL 4	1.034-1.041	L 2	19.6-20.0
LDL IIIB	24.2-24.7	LDL 5	1.041-1.044	L 1	17.7-19.6
LDL IVA	23.3-24.2	LDL 6	1.044-1.051		
LDL IVB	22.0-23.3	LDL 7	1.051-1.060		

Density gradient ultracentrifugation represents the original procedure and is still in use for research, as well as for development and validation of subsequent methods. Although numerous authors consider ultracentrifugation as a "gold standard" procedure for lipoprotein subclasses separation, this technique is less suitable for clinical and epidemiological studies, but is still irreplaceable for preparative purposes. Nondenaturing polyacrylamide gradient gel electrophoresis separates LDL particles on the basis of their charge and size and is the most widely used method. The most recently developed techniques employ the principles of nuclear magnetic resonance spectroscopy (NMR) [18] and ion mobility [19]. Mentioned techniques allow assessment of complete plasma LDL profile, such as particles size, number and concentration. On the other hand, such sophisticated methodologies are very expensive, which limits their wider applicability.

Another drawback has to be emphasised – none of the mentioned techniques is fully interchangeable. Furthermore, due to variety of methodologies and non-uniform nomenclature, an adequate comparison of the results is limited. For instance, Ensign and colleagues [20] recently evaluated correlations between the results of leading methods for LDL subclasses separation based on electrophoretic mobility, ultracentrifugation and NMR and showed a considerable diversity of results - complete agreement among techniques was obtained in 8% of analysed samples [20]. Therefore, any of current techniques cannot be highlighted as superior than others. Each year, several publications present novel technology or improved modifications of existing methods, but additional studies are needed to adequately assess comparability among them. Hopefully, this will identify a reference method, but also the most convenient technique for application in routine laboratory practice.

Atherogenic Properties of Small, Dense LDL Particles

The hypothesis of central role of LDL in atherogenesis survived decades of research. However, after discovering that LDL comprises a mixture of particles with different qualitative characteristics, numerous *in vitro* studies have investigated how such heterogeneity influences the aterogenicity. The results consistently indicated the same: the greatest atherogenic potential belongs to small, dense LDL particles. Over the years, the mechanisms through which small, dense LDL particles achieve such effects have been explained (Table 2).

Table 2. Atherogenic properties of small, dense LDL particles

Cause	Consequence
Reduced affinity for LDL receptors	Longer retention in circulation
Smaller size of the particle	Easier penetration to arterial intima
Enhanced binding to proteoglycans	Longer retention in subendothelium
Reduced content of antioxidants	Accelerated oxidation
Increased affinity for scavanger receptors	Prompt takeover by macrophages

The common denominators for all these processes are increased number of particles and conformational changes in the structure of apoB.

Small, dense LDL particles have a lower affinity for receptors in the liver [21], due to changes in the part of apoB molecule crucial for the binding [22]. In addition, it was reported that conformational changes in other structural components of apoB, although not directly involved in ligand-receptor interactions, also contribute to longer retention of these particles in circulation [23]. On the other hand, small, dense LDL particles have increased affinity for scavanger receptors [24], which can further stimulate foam cell formation and atherogenesis. Also, their smaller size allows easier penetration into subendothelial space [25], whereas reduced amount of sialic acid on their surface enhances binding to proteoglycans, thereby increasing the residence time in subendothelial space [26]. Finally, small, dense LDL particles are more susceptible to oxidation [27], as a consequence of imbalanced amounts of polyunsaturated fatty acids, cholesterol and antioxidants [28, 29]. Bearing in mind high concentration of prooxidants in the subendothelial space, it is reasonable to expect that small, dense LDL particles, with already compromised antioxidant protection, can easily become oxidized. Once formed, oxidized LDL displays its proinflammatory, immunogenic, apoptotic and cytotoxic actions [30]. The spectrum of proinflammatory activities involves hemoatractant effects on circulating monocytes, induction of adhesive molecules expression on endothelial cells and proliferation of monocytes into macrophages [31].

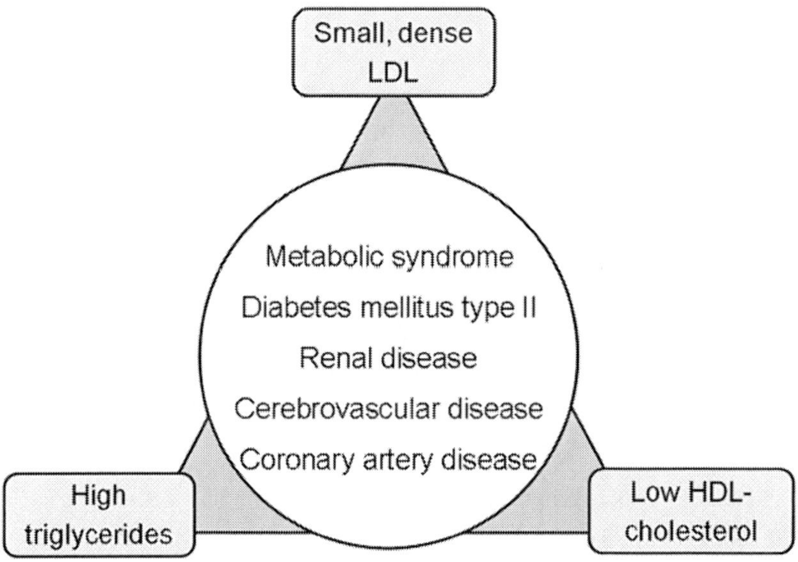

Figure 1. Small, dense LDL particles as a feature of atherogenic lipoprotein phenotype in atherosclerosis-related diseases.

Krauss and colleagues [4] were the first to demonstrate that patients after myocardial infarction have increased small, dense LDL particles. Based on the results of following studies, now is appreciated that subjects with small, dense LDL particles have 3-7 times higher risk of developing CVD [13]. However, it is still not clarified whether small, dense LDL particles are risk factors *per se* or the previously mentioned relation with CVD is just a consequence of broader pathophysiological processes in which they participate [32].

SMALL, DENSE LDL IN ATHEROSCLEROSIS-RELATED DISEASES

Novel evidence indicates that the net effect of dyslipidemia in atherogenesis go beyond plasma LDL-cholesterol concentration [33]. More comprehensive understanding offers the cluster of lipid disturbances named "atherogenic lipoprotein phenotype" or "atherogenic triad" [34]. This term implies concomitant presence of three the most frequent lipid abnormalities in patients with atherosclerosis: increased triglyceride, decreased HDL-cholesterol concentrations and presence of small, dense LDL particles (Figure 1) and its broader application started at the end of 20th century.

Atherogenic lipoprotein phenotype is closely associated with abdominal obesity and insulin resistance and is appreciated as one of the components of the metabolic syndrome [35]. Its clinical importance has been firmly documented, because majority of patients with atherosclerosis-related diseases have this characteristic trait (Figure 1). According to the National Cholesterol Education Program Adult Treatment Panel III (NCEP ATP III) [36], each particular component of this phenotype is atherogenic, but their individual contribution cannot be easily determined. Therefore, some authors readily consider atherogenic lipoprotein phenotype as a whole as a risk factor [37-39].

Development of CVD is a complex process involving oxidative stress, inflammation and dyslipidemia [40]. Based on our previous data in asymptomatic population and in patients with clinically verified atherosclerosis [41-47], we further documented that all of the named risk factors could interact in this pathophysiological mechanism [48-52] and suggested a multimarker approach for the prediction of CVD [53-55]. It has been shown that small, dense LDL subclasses are abundant in patients with coronary atherosclerosis. Up to date, the waste majority of studies recognised small, dense LDL particles as independent predictors of coronary heart disease

[7, 56-73]. On the other hand, the above-mentioned association of LDL particle size with triglyceride and HDL-cholesterol concentrations, as well as with other non-lipid risk factors, such is obesity or insulin resistance [34, 74-76], should not be ignored. These complex metabolic interrelationships further complicate attempts to define intrinsic atherogenic potential of small, dense LDL particles beyond increased LDL-cholesterol concentration. In the majority of aforementioned studies, multivariate logistic regression analysis failed to establish small, dense LDL particles as an independent risk factor for coronary heart disease. Yet, other cross-sectional and prospective studies demonstrated independent contribution of small, dense LDL in the disease development, even after adjustment for other traditional risk factors [7, 59, 61, 63, 67, 70, 71, 73]. In our own study [46], we found an independent association of diminished LDL size and coronary heart disease, regardless of other covariates. However, our further analysis of clinical accuracy revealed that small, dense LDL particles determination provides no additional power than the one obtained by routine lipid status measurement [46]. Taking all together, the question of authentic clinical significance of small, dense LDL particles in prediction of CVD is still under debate. All previously given facts strongly support the necessity of further investigations of LDL heterogeneity, especially if we appreciate that majority of existing studies are dealing with coronary heart disease, while there is a lack of information regarding LDL subclass distribution in other atherosclerosis-related conditions.

Growing evidence suggest that small, dense LDL particles could also be important in the development of ischemic stroke. Because of the high prevalence and mortality rates after stroke [77], effective prevention and treatment are tremendously important goals in clinical practice. Several risk factors have been established for stroke development, among them the most important being age, male sex, hypertension and other cardio- and peripheral-vascular co-morbidity conditions [78-80]. However, there is a lack of information regarding the significance of LDL subclasses determination in this category of patients. Available data showed the association of LDL particle size and subclasses distribution with the development of carotid atherosclerosis [81-86], but similarly as in coronary form, there is insufficient evidence to prove independent contribution of particular subclasses in the onset and progression of stroke. Still, Berneis et al. [73] have shown that LDL size is the most important lipid predictor of carotid intima media thickness, and this predictive potential was retained even after inclusion of all other traditional risk factors. In a Multi-Ethnic Study of Atherosclerosis (MESA) small, dense LDL particles were in strong positive association with intima media thickness of carotid arteries in a mixed population of Caucasians, Chinese, African American and Hispanic population in the United States [87]. Such findings emphasise the importance of starting new studies which should offer more relevant evidence. Our study of LDL size and subclasses distribution in patients with ischemic stroke highlighted small, dense LDL as an independent predictor of both onset and consecutive short-term mortality [88]. These results suggest that LDL subclasses assessment could be useful not only in predicting the risk for the development, but also for identifying patients at higher risk of in-hospital mortality after stroke. In attempting to elucidate possible mechanism through which small, dense LDL particles participate in pathophysiology of stroke, one could rely on previously explained processes in patients with coronary atherosclerosis.

Chronic kidney disease is another condition associated with higher CVD risk [89]. It is also characterised by atherogenic dyslipidemia (Figure 1), mainly as a consequence of decreased removal of triglyceride-rich lipoproteins [89]. Bowden et al. [90] reported that lipoprotein abnormalities, especially altered LDL particle distribution, can be evident even in milder degree of renal impairment. Progressive renal failure is associated with considerable alterations in lipoprotein metabolism, while the greatest risk is observed in patients with end-stage renal disease [91]. However, in end-stage renal disease patients total cholesterol level does not account for the same risk of CVD as in the general population [92, 93]. Still, the relationship between lipoprotein abnormalities and CVD risk in this category of patients remains to be clarified.

A review of literature showed that the type of dyslipidemia in end-stage renal disease can be related to the choice of dialysis modality. Namely, the principal features of renal dyslipidemia in patients on peritoneal dialysis are elevated LDL-cholesterol and preponderance of small, dense LDL particles [94]. On the other side, haemodialysis patients often have total and LDL-cholesterol concentrations within recommended ranges, but elevated triglyceride and reduced HDL-cholesterol concentrations [94]. In accordance, Deighan and colleagues [95] analyzed LDL subfractions and hepatic lipase activity in 75 patients with end-stage renal disease (25 on haemodialysis, 25 on peritoneal dialysis and 25 predialysis patients) to assess the extent of small, dense LDL III subclasses formation in different stages of the disease. The authors [95] found that LDL III particles were increased in all the three patient groups as compared with controls, but the highest proportion was seen among those on peritoneal dialysis. In the same study, plasma triglyceride level and hepatic lipase activity were independent predictors of LDL III abundance [95]. Mekki et al. [96] recently reported that lipid abnormalities in patients undergoing haemodialysis are driven by reduced lipoprotein lipase and hepatic lipase activities, suggesting delayed catabolism of triglyceride-reach lipoproteins. Ambroch et al. [97] further demonstrated that LDL particles in patients with end-stage renal disease are triglyceride-enriched, cholesterol-depleted and smaller in size. In addition, such modified LDL particles were superior substrate for macrophages as compared with fibroblasts [97], indicating a tendency of intracellular accumulation and decreased uptake by the specific receptor pathway.

Results of Yeo et al. [98] indicated that small, dense LDL particles were not associated with CVD in end-stage renal disease patients with angiographically verified coronary atherosclerosis. More recently, Bowden and colleagues [99] published results of the study with more than thousand end-stage renal disease patients. The authors [99] investigated differences between serum lipid parameters and LDL particle number and size in terms of risk stratification when using NCEP ATP III guidelines. In the mentioned study, LDL particle size categorised more patients at increased risk when compared with LDL-cholesterol, non-HDL-cholesterol and triglyceride concentrations and the authors concluded that LDL particles characterisation can be used for advanced screening to predict CVD development [99]. At this point it should be stressed that, in spite of a clear improvement in cardiovascular prevention, the death rate among haemodialysis patients is still high compared to the general population and cardiac causes account for nearly half of overall mortality [100]. Therefore, it is reasonable to anticipate a certain contribution of alterations in LDL subclass distribution to the excess mortality in this category of patients. We recently performed 36 months follow-up study of 122 haemodialysis patients to assess potential predictive role of lipoprotein subclasses on lethal outcome [101]. Our results showed that smaller LDL particles were not more prevalent in the deceased when compared with the survived patients [101], a finding that, in certain manner, confirms previous results of Yeo et al. [98]. Nevertheless, further prospective studies should reveal the importance of LDL particles characterisation in attempting to improve survival of the patients until renal transplantation.

Renal transplant recipients also have an increased risk of CVD development [102]. Similarly to pre-transplantation period, the patients still suffer from dyslipidemia [103]. As documented by Rajman et al. [104], the presence of small, dense LDL particles is an early feature of the uremic dyslipidemia and might persists even after renal transplantation. Side effects of immunosuppressive medications represent significant sources of post-transplant dyslipidemia. It has been shown that cyclosporine administration is more likely associated with dyslipidemia than the use of tacrolimus [105]. Recently, we have investigated heterogeneity of LDL particles and the influence of immunosuppressive drugs on their distribution in pediatric renal transplant recipients (unpublished data). We found that pediatric renal transplant recipients have significantly higher proportion of small, dense LDL particles than controls, as well as a shift towards smaller particles in patients treated with cyclosporine when compared with subjects on tacrolimus. In the study by Apanay et al. [106], patients treated with higher cyclosporine levels showed significantly higher LDL oxidability when compared to controls. Furthermore, Cofan et al. [107] noted that tacrolimus therapy resulted in a LDL oxidation profile similar to that in healthy controls. It appears that conversion from cyclosporine to tacrolimus might have beneficial effects on both LDL particles distribution and susceptibility to oxidation, thereby reducing LDL's atherogenic potential. Overall, these findings made it obvious that pre-transplant condition or dialysis-specific alterations in LDL subclass distribution can be continued in post-transplant period, as it was documented by Rajman et al. [104]. It is also clear that traditional lipids and apolipoproteins measurements could not be sufficient for accurate and timely estimation of atherogenic dyslipidemia in end-stage renal disease, as well as for post-transplantation patients. The usefulness of advanced lipid testing could be particularly true for small, dense LDL particles estimation.

Based on all previously discussed data about the role of qualitative and quantitative characteristics of LDL subclasses in the development of atherosclerosis-related diseases, this chapter should also briefly address pharmacological influences on LDL particles size distribution. Among available hypolipemic agents, fibrates seem to possess the most prominent beneficial effect [108]. It has also been demonstrated that statins could reduce the proportion of small, dense LDL particles, but with significant variations among different members in this group [109-112]. Nowadays, research is extended toward exploring new therapeutic options beyond ordinarily prescribed hypolipemics [113, 114]. One could presume that future therapeutic approach might shift from prevailing use of statins towards drugs that, besides correction of LDL-cholesterol level, can efficiently improve disturbances in LDL subclasses distribution, too.

PERSPECTIVES IN ROUTINE LABORATORY DIAGNOSTICS

The basic criteria for routine use of each potential new CVD risk marker are firmly established [115]. First of all, it is required to satisfy exceptionally strict rules for the characteristics of analytical technique (regarding availability, simplicity and standardisation), as well as for laboratory costs. Next, the predictive power of a new marker has to be confirmed in prospective studies, whereby it needs to be stronger than prognostic value of existing traditional risk factors. Finally, it is equally important that new marker can be applied for risk prediction in all populations.

As it has been illustrated above, the function of small, dense LDL particles in the development of atherosclerosis is generally well understood. In spite of that, clinical significance of their assessment is still controversial. The key question that remains to be answered is whether specific LDL subclasses are independent risk factors for atherosclerosis-related diseases. If so, presence of small, dense LDL particles should not be neglected in general risk assessment. The opposing alternative is that relative risk attributable to small, dense LDL subclasses is merely a consequence of complex interactions with other traditional risk factors. In general, to indisputably determine independent association of small, dense LDL with CVD, confirmative findings of multiple logistic regression analysis are obligatory. Although many cross-sectional and prospective studies have dealt with this issue, their results are not homogenous.

In 2009, American Association for Clinical Chemistry (AACC) published recommendations for laboratory assessment of novel risk factors in primary prevention of CVD and stroke [116]. Analysing available data and following American College of Cardiology (ACC) and American Heart Association (AHA) guidelines for implementation of new diagnostic or therapeutic procedures [117], Wilson et al. [116] defined recommendations for determining LDL size and subclasses distribution. The authors concluded that, unlike for several other proposed CVD risk factors, none of the above mentioned criteria have been still satisfied in the case of LDL subclasses (Table 3).

In the introducing section of the above mentioned publication, it was emphasised that the recommendations were based on the results of prospective studies in healthy populations published until February 2005. Neither retrospective, nor anamnestic studies were analysed, except for estimation of marker's significance in secondary prevention [118].

Table 3. Guidelines for the measurement of LDL subclasses in primary prevention

American Association for Clinical Chemistry recommendations [116]	
Recommendation 1	Data analyses of existing studies are generally not adequate to show added benefit over standard risk assessment for primary prevention.
Recommendation 2	There are insufficient data that measurement of lipoprotein subclasses over time is useful to evaluate the effects of treatments.
Recommendation 3	Several methods are available to assess lipoprotein subclasses. Standardization is needed for this technology.

Analysis of Wilson and associates [116] did not include studies on clinical significance of small, dense LDL-cholesterol concentration (sdLDL-cholesterol), measured by a method based on selective precipitation in the presence of heparin and Mg^{2+} ions [119]. Preliminary results on this new parameter were published in 2003 [119], but meanwhile the authors modified and improved the existing method [120], examined preanalytical variations [121, 122] and clinical significance [123-125]. This group further demonstrated that increased concentration of sdLDL-cholesterol is associated with the extent of coronary [123] and carotid atherosclerosis [126], but also that statin and fibrate therapy efficiently lowers sdLDL-cholesterol level in diabetics [127]. Having in mind the simplicity and availability of this proposed method, its value for prediction of the disease and evaluation of the therapy, it seems reasonable to expect that this parameter would eventually fulfill the ACC/AHA criteria [117] and would be recommended for routine use in primary prevention of CVD.

CONCLUSION

The distribution of LDL subclasses undoubtedly undergoes major changes in patients with atherosclerosis, but clinical significance of their assessment is still unresolved. Small, dense LDL particles seem to be important predictors of progression of atherosclerosis-related diseases and their predominance have been accepted as an emerging cardiovascular risk factor by the NCEP ATP III [36]. The studies discussed above suggest that screening for the presence of small, dense LDL may contribute in directing specific interventions for cardiovascular prevention. Based on current experiences, possible indications for LDL subfraction assessment can be summarized as follows: management of any form of atherosclerotic disease, diagnostic evaluation of primary and secondary dyslipidemias, as well as follow-up of asymptomatic individuals that are at increased risk for future development of atherosclerosis. Is the latest research on LDL particle heterogeneity sufficient enough to provide evidence that can ensure their implementation in clinical practice? This question is waiting to be answered in the upcoming NCEP ATP IV guidelines.

ACKNOWLEDGEMENTS

This work was supported by a grant from the Ministry of Education and Science, Republic of Serbia (Project No. 175035).

REFERENCES

[1] Fredrickson, DS; Levy, RI; Lees, RS. Fat transport in lipoproteins - an integrated approach to mechanisms and disorders. *N. Engl. J. Med.,* 1967, 276, 148-156.

[2] Gofman, JW; Lindgren, F. The role of lipids and lipoproteins in atherosclerosis. *Science,* 1950, 111, 166-171.

[3] Hevonoja, T; Pentikäinen, MO; Hyvönen, MT; Kovanen, PT; Ala-Korpela, M. Structure of low density lipoprotein (LDL) particles: basis for understanding molecular changes in modified LDL. *Biochim. Biophys. Acta,* 2000, 1488, 189-210.

[4] Krauss, RM; Burke, DJ. Identification of multiple subclasses of plasma low density lipoproteins in normal humans. *J. Lipid. Res.,* 1982, 23, 97-104.

[5] Campos, H; Blijlevens, E; McNamara, JR; Ordovas, JM; Posner, BM; Wilson, PW; Castelli, WP; Schaefer, EJ. Low density lipoprotein particle size distribution: results from the Framingham Offspring Study. *Arterioscler. Thromb.,* 1992, 12, 1410-1419.

[6] Packard, CJ; Gaw, A; Demant, T; Shepherd J. Development and application of a multicompartmental model to study very low density lipoprotein subfraction metabolism. *J. Lipid. Res.,* 1995, 36, 172-187.

[7] Zambon, A; Hokanson, JE; Brown, BG; Brunzell, JD. Evidence for a new pathophysiological mechanism for coronary artery disease regression: hepatic lipase-mediated changes in LDL density. *Circulation,* 1999, 99, 1959-1964.
[8] Guerin, M; Le Goff, W; Lassel, TS; Van Tol, A; Steiner, G; Chapman, MJ. Atherogenic role of elevated CE transfer from HDL to VLDL(1) and dense LDL in type 2 diabetes: impact of the degree of triglyceridemia. *Arterioscler. Thromb. Vasc. Biol.,* 2001, 21, 282-288.
[9] Yamada, ND; Shames, M; Takahashi, K; Havel, RJ. Metabolism of apolipoprotein B-100 in large very low density lipoproteins in blood plasma. *J. Clin. Invest.,* 1988, 82, 2106-2113.
[10] Jansen, H; Hop, W; van Tol, A; Bruschke, AV; Birkenhager, JC. Hepatic lipase and lipoprotein lipase are not major determinants of the low density lipoprotein subclass pattern in human subjects with coronary heart disease. *Atherosclerosis,* 1994, 107, 45-54.
[11] Deckelbaum, RJ; Granot, E; Oschry, Y; Rose, L; Eisenberg, S. Plasma triglyceride determines structure-composition in low and high density lipoproteins. *Arteriosclerosis,* 1984, 4, 225-231.
[12] Austin, MA; Krauss, RM. Genetic control of low-density lipoprotein subclasses. *Lancet*, 1986, 2, 592-595.
[13] Berneis, KK; Krauss, RM. Metabolic origins and clinical significance of LDL heterogeneity. *J. Lipid. Res.,* 2000, 43, 1363-1379.
[14] Rizzo, M; Berneis, K. Low-density lipoprotein size and cardiovascular risk assessment. *QJM,* 2006, 99, 1-14.
[15] Dreon, DM; Fernstrom, HA; Williams, PT; Krauss, RM. LDL subclass patterns and lipoprotein response to a low-fat, high-carbohydrate diet in women. *Arterioscler. Thromb. Vasc. Biol.,* 1997, 17, 707-714.
[16] Mukamal, KJ; Mackey, RH; Kuller, LH; Tracy, RP; Kronmal, RA; Mittleman, MA; Siscovick, DS. Alcohol consumption and lipoprotein subclasses in older adults. *J. Clin. Endocrinol. Metab.,* 2007, 92, 2559-2566.
[17] de Graaf, J; Swinkels, DW; Demacker, PN; de Haan, AF; Stalenhoef, AF. Differences in the low density lipoprotein subfraction profile between oral contraceptive users and controls. *J. Clin. Endocrinol. Metab.*, 1993, 76, 197-202.
[18] Otvos, JD. Measurement of lipoprotein subclass profiles by nuclear magnetic resonance spectroscopy. In: Rifai N, Warnick GR, Dominiczak MH, editors. Handbook of Lipoprotein Testing. Washington, DC: AACC Press; 2000; 609-623.

[19] Caulfield, MP; Li, S; Lee, G; Blanche, PJ; Salameh, WA; Benner, WH; Reitz, RE; Krauss, RM. Direct determination of lipoprotein particle sizes and concentrations by ion mobility analysis. *Clin. Chem.,* 2008, 54, 1307-1316.

[20] Ensign, W; Hill, N; Heward, CB. Disparate LDL phenotypic classification among 4 different methods assessing LDL particle characteristics. *Clin. Chem.,* 2006, 52, 1722-1727.

[21] Galeano, NF; Milne, R; Marcel, YL; Walsh, MT; Levy, E; Ngu'yen, TD; Gleeson, A; Arad, Y; Witte, L; Al-Haideri, M; Deckelbaum, RJ. Apoprotein B structure and receptor recognition of triglyceride-rich low density lipoprotein LDL is modified in small LDL but not in triglyceride-rich LDL of normal size. *J. Biol. Chem.*, 1994, 7, 511-519.

[22] Teng, B; Sniderman, AD; Krauss, RM; Kwiterovich, PO; Milne, RW; Marcel, YL. Modulation of apolipoprotein B antigenic determinants in human low density lipoprotein subclasses. *J. Biol. Chem.*, 1985, 260, 5067-5072.

[23] Chen, GC; Liu, W; Duchateau, P; Allaart, J; Hamilton, RL; Mendel, CM; Lau, K; Hardman, DA; Frost, PH; Malloy, MJ; Kane, JP. Conformational differences in human apolipoprotein B-100 among subspecies of low density lipoproteins (LDL). Association of altered proteolytic accessibility with decreased receptor binding of LDL subspecies from hypertriglyceridemic subjects. *J. Biol. Chem.*, 1994, 269, 29121-29128.

[24] Galeano, NF; Al-Haideri, M; Keyserman, F; Rumsey, SC; Deckelbaum, RJ. Small dense low density lipoprotein has increased affinity for LDL receptor-independent cell surface binding sites: a potential mechanism for increased atherogenicity. *J. Lipid. Res.,* 1998, 39, 1263-1273.

[25] Nordestgaard, BG; Zilversmit, DB. Comparison of arterial intimal clearances of LDL from diabetic and nondiabetic cholesterol-fed rabbits. Differences in intimal clearance explained by size differences. *Arteriosclerosis,* 1989, 9, 176-183.

[26] Hurt-Camejo, E; Camejo, G; Rosengren, B; Lopez, F; Wiklund, O; Bondjers, G. Differential uptake of proteoglycan-selected subfractions of low density lipoprotein by human macrophages. *J. Lipid. Res.,* 1990, 31, 1387-1398.

[27] de Graaf, J; Hak-Lemmers, HL; Hectors, MP; Demacker, PN; Hendriks, JC; Stalenhoef, AF. Enhanced susceptibility to *in vitro* oxidation of the dense low density lipoprotein subfraction in healthy subjects. *Arterioscler. Thromb.*, 1991, 11, 298-306.

[28] Tribble, DL; Holl, LG; Wood, PD; Krauss, RM. Variations in oxidative susceptibility among six low density lipoprotein subfractions of differing density and particle size. *Atherosclerosis,* 1992, 93, 89-99.

[29] Tribble, DL; Rizzo, M; Chair, A; Lewis, DM; Blanche, PJ; Krauss, RM. Enhanced oxidative susceptibility and reduced antioxidant content of metabolic precursors of small, dense low-density lipoproteins. *Am. J. Med.,* 2001, 110, 103-110.

[30] Chisolm, GM; Steinberg, D. The oxidative modification hypothesis of atherogenesis: an overview. *Free Radic. Biol. Med.,* 2000, 28, 1815-1826.

[31] Vekic, J; Kotur-Stavuljevic, J; Zeljkovic, A; Stefanovic, A; Bogavac-Stanojevic, N; Vujovic, A. Oxidative stress, HDL and atherosclerosis. *Anti-Inflammatory Anti-Allergy Agents Med. Chem.*, 2008, 7, 158-165.

[32] Rizzo M; Berneis K; Zeljkovic A; Vekic J. Should we routinely measure low-density and high-density lipoprotein subclasses? *Clin. Lab.,* 2009, 55, 421-429.

[33] Rizzo, M; Berneis, K. Lipid triad or atherogenic lipoprotein phenotype: a role in cardiovascular prevention? *J. Atheroscler. Thromb.*, 2005, 12, 237-139.

[34] Austin, MA; King, MC; Vranizan, KM; Krauss, RM. Atherogenic lipoprotein phenotype. A proposed genetic marker for coronary heart disease risk. *Circulation*, 1990, 82, 495-506.

[35] Rizzo, M; Berneis, K. Small, dense low-density-lipoproteins and the metabolic syndrome. *Diabetes Metab. Res.,* 2007, 23, 14-20.

[36] National Cholesterol Education Program (NCEP). Expert Panel on Detection, Evaluation, and Treatment of High Blood Cholesterol in Adults (Adult Treatment Panel III). Third Report of the National Cholesterol Education Program (NCEP) Expert Panel on Detection, Evaluation, and Treatment of High Blood Cholesterol in Adults (Adult Treatment Panel III) final report. *Circulation,* 2002, 106, 3143-3421.

[37] Rizzo, M; Pernice, V; Frasheri, A; Berneis, K. Atherogenic lipoprotein phenotype and LDL size and subclasses in patients with peripheral arterial disease. *Atherosclerosis*, 2008, 197, 237-241.

[38] Berneis, K; Rizzo, M; Fruzzetti, F; Lazzaroni, V; Carmina, E. Atherogenic lipoprotein phenotype and low-density lipoproteins size and subclasses in women with polycystic ovary syndrome. *J. Clin. Endocrinol. Metab.,* 2007, 92, 186-189.

[39] Rizzo, M; Trepp, R; Berneis, K; Christ, ER. Atherogenic lipoprotein phenotype and LDL size and subclasses in patients with growth hormone

deficiency before and after short-term replacement therapy. *Eur. J. Endocr.,* 2007, 156, 361-367.

[40] Rizzo, M; Kotur-Stevuljevic, J; Berneis, K; Spinas, G; Rini, GB; Jelic-Ivanovic, Z; Spasojevic-Kalimanovska, V; Vekic, J. Atherogenic dyslipidemia and oxidative stress: a new look. *Transl. Res.,* 2009, 153, 217-223.

[41] Vekic, J; Zeljkovic, A; Jelic-Ivanovic, Z; Spasojevic-Kalimanovska, V; Bogavac-Stanojevic, N; Memon, L; Spasic, S. Small, dense LDL cholesterol and apolipoprotein B: relationship with serum lipids and LDL size. *Atherosclerosis*, 2009, 207, 496-501.

[42] Vekic, J; Topic, A; Zeljkovic, A; Jelic-Ivanovic, Z; Spasojevic-Kalimanovska, V. LDL and HDL subclasses and their relationship with Framingham risk score in middle-aged Serbian population. *Clin. Biochem.*, 2007, 40, 310-316.

[43] Memon, L; Spasojevic-Kalimanovska, V; Bogavac-Stanojevic, N; Kalimanovska-Ostric, D; Jelic-Ivanovic, Z; Spasic, S; Topic, A. Association of C-reactive protein with the presence and extent of angiographically verified coronary artery disease. *Tohoku J. Exp. Med.*, 2006, 209, 197-206.

[44] Kotur-Stevuljevic, J; Spasic, S; Stefanovic, A; Zeljkovic, A; Bogavac-Stanojevic, N; Kalimanovska-Ostric, D; Spasojevic-Kalimanovska, V; Jelic-Ivanovic, Z. Paraoxonase-1 (PON1) activity, but not $PON1_{Q192R}$ phenotype, is a predictor of coronary artery disease in a middle-aged Serbian population. *Clin. Chem. Lab. Med.*, 2006, 44, 1106-1113.

[45] Jelic-Ivanovic, Z; Memon, L; Spasojevic-Kalimanovska, V; Bogavac-Stanojevic, N; Spasic, S. Independent association of high serum uric acid concentration with angiographically defined coronary artery disease. *Tohoku J. Exp. Med.,* 2007, 211, 369-377.

[46] Zeljkovic, A; Spasojevic-Kalimanovska, V; Vekic, J; Jelic-Ivanovic, Z; Topic, A; Bogavac-Stanojevic, N; Spasic, S; Vujovic, A; Kalimanovska-Ostric, D. Does simultaneous determination of LDL and HDL particle size improve prediction of coronary artery disease risk? *Clin. Exp. Med.*, 2008, 8, 109-116.

[47] Jelic-Ivanovic, Z; Bujisic, N; Spasic, S; Bogavac-Stanojevic, N; Spasojevic-Kalimanovska, V; Kotur-Stevuljevic, J. Circulating sTWEAK improves the prediction of coronary artery disease. *Clin. Biochem.*, 2009, 42, 1381-1386.

[48] Kotur-Stevuljevic, J; Memon, L; Stefanovic, A; Spasic, S; Spasojevic-Kalimanovska, V; Bogavac-Stanojevic, N; Kalimanovska-Ostric, D;

Jelic-Ivanovic, Z; Zunic, G. Correlation of oxidative stress parameters and inflammatory markers in coronary artery disease patients. *Clin. Biochem.,* 2007, 40, 181-187.

[49] Vekic, J; Kotur-Stevuljevic, J; Jelic-Ivanovic, Z; Spasic, S; Spasojevic-Kalimanovska, V; Topic, A; Zeljkovic, A; Stefanovic, A; Zunic, G. Association of oxidative stress and PON1 with LDL and HDL particle size in middle-aged subjects. *Eur. J. Clin. Invest.,* 2007, 37, 715-723.

[50] Kotur-Stevuljevic, J; Spasic, S; Jelic-Ivanovic, Z; Spasojevic-Kalimanovska, V; Stefanovic, A; Vujovic, A; Memon, L; Kalimanovska-Ostric, D. PON1 status in influenced by oxidative stress and inflammation in coronary heart disease patients. *Clin. Biochem.*, 2008, 41, 1067-1073.

[51] Zeljkovic, A; Bogavac-Stanojevic, N; Jelic-Ivanovic, Z; Spasojevic-Kalimanovska, V; Vekic, J; Spasic, S. Combined effects of small apolipoprotein (a) isoforms and small, dense LDL on coronary artery disease risk. *Arch. Med. Res.*, 2009, 40, 29-35.

[52] Vekic, J; Jelic-Ivanovic, Z; Spasojevic-Kalimanovska, V; Memon, L; Zeljkovic, A; Bogavac-Stanojevic, N; Spasic, S. High serum uric acid and low-grade inflammation are associated with smaller LDL and HDL particles. *Atherosclerosis,* 2009, 203, 236-242.

[53] Bogavac-Stanojevic, N; Jelic-Ivanovic, Z; Spasojevic-Kalimanovska, V; Spasic, S; Kalimanovska-Ostric, D. Lipid and inflammatory markers for the prediction of coronary artery disease: a multi-marker approach. *Clin. Biochem.,* 2007, 40, 1000-1006.

[54] Bogavac-Stanojevic, N; Ivanova Petrova, G; Jelic-Ivanovic, Z; Memon, L; Spasic, S. Cost-effectiveness analysis in diagnosis of coronary artery disease: choice of laboratory markers. *Clin. Biochem.,* 2007, 40, 1180-1187.

[55] Lakic, D; Bogavac-Stanojevic, N; Jelic-Ivanovic, Z; Kotur-Stevuljevic, J; Spasic, S; Kos, M. A multi-marker approach for the prediction of coronary artery disease: cost-effectivness analysis. *Value Health,* 2010, 13, 770-777.

[56] Austin, MA; Breslow, JL; Hennekens, CH; Buring, JE; Willett, WC; Krauss, RM. Low-density lipoprotein subclass patterns and risk of myocardial infarction. *JAMA,* 1988, 260, 1917-1921.

[57] Campos, H; Genest, J; Blijlevens, E; McNamara, JR; Jenner, JL; Ordovas, JM; Wilson, PW; Schaefer, EJ. Low density lipoprotein particle size and coronary artery disease. *Arteriosclerosis,* 1992, 12, 187-195.

[58] Coresh, J; Kwiterovich, PO; Smith, HH; Bachorik, PS. Association of plasma triglyceride concentration and LDL particle diameter, density, and chemical composition with premature coronary artery disease in men and women. *J. Lipid. Res.,* 1993, 34, 1687-1697.
[59] Griffin, BA; Freeman, DJ; Tait, GW; Thomson, J; Caslake, MJ; Packard, CJ; Shepherd, J. Role of plasma triglyceride in the regulation of plasma low density lipoprotein (LDL) subfractions: relative contribution of small dense LDL to coronary heart disease risk. *Atherosclerosis,* 1994, 106, 241-253.
[60] Stampfer, MJ; Krauss, RM; Ma, J; Blanche, PJ; Holl, LG; Sacks, FM; Hennekens, CH. A study of triglyceride level, low-density lipoprotein particle diameter, and risk of myocardial infarction. *JAMA*, 1996, 276, 882-888.
[61] Gardner, CD; Fortmann, SP; Krauss, RM. Association of small low-density lipoprotein particles with the incidence of coronary artery disease in men and women. *JAMA*, 1996, 276, 875-881.
[62] Mack, WJ; Krauss, RM; Hodis, HN. Lipoprotein subclasses in the Monitored Atherosclerosis Regression Study (MARS). Treatment effects and relation to coronary angiographic progression. *Arterioscler. Thromb. Vasc. Biol.,* 1996, 16, 697-704.
[63] Lamarche, B; Tchernof, A; Moorjani, S; Cantin, B; Dagenais, GR; Lupien, PJ; Despres, JP. Small, dense low density lipoprotein particles as a predictor of the risk of ischemic heart disease in men. P results from the Quebec Cardiovascular Study. *Circulation*, 1997, 95, 69-75.
[64] Erbey, JR; Robbins, D; Forrest, KY; Orchard TJ. Low-density lipoprotein particle size and coronary artery disease in a childhood-onset type 1 diabetes population. *Metabolism*, 1999, 48, 531-534.
[65] Austin, MA; Rodriguez, BL; McKnight, B; McNeely, MJ; Edwards, KL; Curb, JD; Sharp, DS. Low-density lipoprotein particle size, triglycerides, and high density lipoprotein cholesterol as risk factors for coronary heart disease in older Japanese-American men. *Am. J. Cardiol.,* 2000, 86, 412-416.
[66] Kamigaki, AS; Siscovick, DS; Schwartz, SM; Psaty, BM; Edwards, KL; Raghunathan, TE; Austin, MA. Low density lipoprotein particle size and risk of early-onset myocardial infarction in women. *Am. J. Epidemiol.*, 2001, 153, 939-945.
[67] Rosenson, RS; Otvos, JD; Freedman, DS. Relations of lipoprotein subclass levels and low-density lipoprotein size to progression of coronary artery disease in the Pravastatin Limitation of Atherosclerosis

in the Coronary Arteries (PLAC-I) trial. *Am. J. Cardiol.,* 2002, 90, 89-94.

[68] Blake, GJ; Otvos, JD; Rifai, N; Ridker, PM. Low-density lipoprotein particle concentration and size as determined by nuclear magnetic resonance spectroscopy as predictors of cardiovascular disease in women. *Circulation,* 2002, 106, 1930-1937.

[69] Kuller, L; Arnold, A; Tracy, R; Otvos, J; Burke, G; Psaty, B; Siscovick, D; Freedman, DS; Kronmal, R. Nuclear magnetic resonance spectroscopy of lipoproteins and risk of coronary heart disease in the cardiovascular health study. *Arterioscler. Thromb. Vasc. Biol.,* 2002, 22, 1175-1780.

[70] Koba, S; Hirano, T; Kondo, T; Shibata, M; Suzuki, H; Murakami, M; Geshi, E; Katagiri, T. Significance of small dense low-density lipoproteins and other risk factors in patients with various types of coronary heart disease. *Am. Heart J.,* 2002, 144, 1026-1035.

[71] Yoon, Y; Song, J; Park, HD; Park, KU; Kim, JQ. Significance of small dense low-density lipoproteins as coronary risk factor in diabetic and non-diabetic Korean populations. *Clin. Chem. Lab. Med.*, 2005, 43, 431-437.

[72] St-Pierre, AC; Cantin, B; Dagenais, GR; Mauriege, P; Bernard, PM; Despres, JP; Lamarche, B. Low-density lipoprotein subfractions and the long-term risk of ischemic heart disease in men: 13-year follow-up data from the Quebec Cardiovascular Study. *Arterioscler. Thromb. Vasc. Biol.,* 2005, 25, 553-559.

[73] Berneis, K; Jeanneret, C; Muser, J; Felix, B; Miserez, AR. Low density lipoprotein size and subclasses are markers of clinically apparent and non-apparent atherosclerosis in type 2 diabetes. *Metabolism,* 2005, 54, 227-234.

[74] McNamara, JR; Campos, H; Ordovas, JM; Peterson, J; Wilson, PWF; Schaefer, EJ. Effect of gender, age, and lipid status on low density lipoprotein subfraction distribution. Results of the Framingham Offspring Study. *Arteriosclerosis*, 1987, 7, 483-490.

[75] Reaven, GM; Chen, YD; Jeppesen, J; Maheux, P; Krauss, RM. Insulin resistance and hyperinsulinemia in individuals with small, dense low density lipoprotein particles. *J. Clin. Invest.,* 1993, 92, 141-146.

[76] Hulthe, J; Bokemark, L; Wikstrand, J; Fagerberg, B. The metabolic syndrome, LDL particle size, and atherosclerosis: the Atherosclerosis and Insulin Resistance (AIR) study. *Arterioscler. Thromb. Vasc. Biol.,* 2000, 20, 2140-2147.

[77] Murray, CJ; Lopez, AD. Mortality by cause for eight regions of the world: global burden of disease study. *Lancet*, 1997, 349, 1269-1276.

[78] Kirshner, HS. Differentiating ischemic stroke subtypes: risk factors and secondary prevention. *J. Neurol. Sci.*, 2009, 279, 1-8.

[79] Ohira, T; Shahar, E; Chambless, LE; Rosamond, WD; Mosley, TH Jr; Folsom, AR. Risk factors for ischemic stroke subtypes: the Atherosclerosis Risk in Communities study. *Stroke*, 2006, 37, 2493-2498.

[80] Folsom, AR; Rosamond, WD; Shahar, E; Cooper, LS; Aleksic, N; Nieto, FJ; Rasmussen, ML; Wu, KK. Prospective study of markers of hemostatic function with risk of ischemic stroke. The Atherosclerosis Risk in Communities (ARIC) Study Investigators. *Circulation,* 1999, 100, 736-742.

[81] Landray, MJ; Sagar, G; Muskin, J; Murray, S; Holder, RL; Lip, GYH. Association of atherogenic low-density lipoprotein subfractions with carotid atherosclerosis. *QJM,* 1998, 91, 345-351.

[82] Skoglund-Andersson, C; Tang, R; Bond, MG; de Faire, U; Hamsten, A; Karpe, F. LDL particle size distribution is associated with carotid intima-media thickness in healthy 50-year-old men. *Arterioscler. Thromb. Vasc. Biol.*, 1999, 19, 2422-2430.

[83] Liu, ML; Ylitalo, K; Nuotio, I; Salonen, R; Salonen, JT; Taskinen, MR. Association between carotid intima-media thickness and low-density lipoprotein size and susceptibility of low-density lipoprotein to oxidation in asymptomatic members of familial combined hyperlipidemia families. *Stroke,* 2002, 33, 1255-1260.

[84] Slowik, A; Iskra, T; Turaj, W; Hartwich, J; Dembinska-Kiec, A; Szczudlik, A. LDL phenotype B and other lipid abnormalities in patients with large vessel disease and small vessel disease. *Neurol. Sci.*, 2003, 214, 11-16.

[85] Watanabe, T; Koba, S; Kawamura, M; Itokawa, M; Idei, T; Nakagawa, Y; Iguchi, T; Katagiri, T. Small dense low-density lipoprotein and carotid atherosclerosis in relation to vascular dementia. *Metabolism,* 2004, 53, 476-482.

[86] Inukai, T; Yamamoto, R; Suetsugu, M; Matsumoto, S; Wakabayashi, S; Inukai, Y; Matsutomo, R; Takebayashi, K; Aso, Y. Small low-density lipoprotein and small low-density lipoprotein/total low-density lipoprotein are closely associated with intima-media thickness of the carotid artery in Type 2 diabetic patients. *J. Diabetes Complications*, 2005, 19, 269-275.

[87] Mora, S; Szklo, M; Otvos, JD; Greenland, P; Psaty, BM; Goff, DC Jr; O'Leary, DH; Saad, MF; Tsai, MY; Sharrett, AR. LDL particle subclasses, LDL particle size, and carotid atherosclerosis in the Multi-Ethnic Study of Atherosclerosis (MESA). *Atherosclerosis,* 2007, 192, 211-217.

[88] Zeljkovic, A; Vekic, J; Spasojevic-Kalimanovska, V; Jelic-Ivanovic, Z; Bogavac-Stanojevic, N; Gulan, B; Spasic, S. LDL and HDL subclasses in acute ishemic stroke: prediction of risk and short-term mortality. *Atherosclerosis,* 2010, 210, 548-554.

[89] Kaysen, GA. Metabolic syndrome and renal failure: similarities and differences. *Panminerva Med.*, 2006, 48, 151-164.

[90] Bowden, RG; Hebert, S; Wilson, R; Gentile, M, Lanning, BA. Comparison of lipid measures and risk stratification among end-stage renal disease patients. *J. Nephrol.*, 2007, 20, 212-218.

[91] Sarnak, MJ; Levey, AS; Schoolwerth, AC; Coresh, J; Culleton, B; Hamm, LL; McCullough, PA; Kasiske, BL; Kelepouris, E; Klag, MJ; Parfrey, P; Pfeffer, M; Raij, L; Spinosa, DJ; Wilson, PW. Kidney disease as a risk factor for development of cardiovascular disease: a statement from the American Heart Association Councils on Kidney in Cardiovascular Disease, High Blood Pressure Research, Clinical Cardiology, and Epidemiology and Prevention. *Circulation*, 2003, 108, 2154-2169.

[92] Fleischmann EH; Bower JD; Salahudeen AK. Risk factor paradox in hemodialysis: better nutrition as a partial explanation. *ASAIO Journal*, 2001, 47, 74-81.

[93] Iseki K; Yamazato M; Tozawa M; Takishita S. Hypocholesterolemia is a significant predictor of death in a cohort of chronic hemodialysis patients. *Kidney Int.*, 2002, 61, 1887-1893.

[94] Prichard, SS. Impact of dyslipidemia in end-stage renal disease. *J. Am. Soc. Nephrol.*, 2003, 14, S315-S320.

[95] Deighan, CJ; Caslake, MJ; McConnell, M; Boulton-Jones, JM; Packard,CJ. Atherogenic lipoprotein phenotype in end-stage renal failure: origin and extent of small dense low-density lipoprotein formation. *Am. J. Kidney Dis.*, 2000, 35, 852-862.

[96] Mekki, K; Prost, J; Remaoun, M; Belleville, J; Bouchenak, M. Long term hemodialysis aggravates lipolytic activity reduction and very low density, low density lipoproteins composition in chronic renal failure patients. *BMC Cardiovasc. Disord,* 2009, 9, 41.

[97] Ambrosch, A; Domroese, U; Westphal, S; Dierkes, J; Augustin, W; Neumann, KH; Luley, C. Compositional and functional changes of low-density lipoprotein during hemodialysis in patients with ESRD. *Kidney Int.,* 1998, 54, 608-617.

[98] Yeo, Y; Byun, SW; Lee, JY; Min, WK; Park, JS; Kim, SB. Lack of association between small dense low-density lipoprotein levels and coronary artery disease in chronic hemodialysis patients. *Am. J. Nephrol.*, 2009, 30, 310-314.

[99] Bowden, RG; Wilson, RL; Beaujean, AA. LDL particle size and number compared with LDL cholesterol and risk categorization in end-stage renal disease patients. *J. Nephrol.,* 2011, doi: 10.5301/JN.2011.6376.

[100] Herzog, CA; Ma, JZ; Collins, AJ. Poor long-term survival after acute myocardial infarction among patients on long-term dialysis. *N. Engl. J. Med.,* 1998, 339, 799-805.

[101] Vekic, J; Zeljkovic, A; Bogavac-Stanojevic, N; Jelic-Ivanovic, Z; Spasojevic-Kalimanovska, V; Simic-Ogrizovic, S; Dopsaj, V; Spasic, S. Cox proportional hazard model analysis of survival in end-stage renal disease patients with small-sized high-density lipoprotein particles. *Clin. Biochem.,* 2011, 44, 635-641.

[102] Kasiske, BL. Cardiovascular disease after renal transplantation. *Semin. Nephrol.,* 2000, 20, 176-187.

[103] Vaziri, ND. Causes of dysregulation of lipid metabolism in chronic renal failure. *Semin. Dial.,* 2009, 22, 644-651.

[104] Rajman, I; Harper, L; McPake, D; Kendall, MJ; Wheeler, DC. Low-density lipoprotein subfraction profiles in chronic renal failure. *Nephrol. Dial. Transplant.,* 1998, 13, 2281-2287.

[105] Baid-Agrawal, S; Delmonico, FL; Tolkoff-Rubin, NE; Farrell, M; Williams, WW; Shih, V; Auchincloss, H; Cosimi, AB; Pascual, M. Cardiovascular risk profile after conversion from cyclosporine A to tacrolimus in stable renal transplant recipients. *Transplantation,* 2004, 77, 1199-1202.

[106] Apanay, DC; Neylan, JF; Ragab, MS; Sgoutas, DS. Cyclosporine increases the oxidizability of low-density lipoproteins in renal transplant recipients. *Transplantation,* 1994, 5, 663-669.

[107] Cofan, F; Zambon, D; Laguna, JC; Ros, E; Casals, E; Cofan, M; Campistol, JM; Oppenheimer, F. Oxidation of low-density lipoproteins in renal transplant recipients treated with tacrolimus. *Transplantation Proc.*, 2002, 34, 377-378.

[108] Rizzo, M; Berneis, K. The clinical significance of the size of low density lipoproteins and the modulation of subclasses by fibrates. *Curr. Med. Res. Opin.,* 2007; 23:1103-1111.

[109] Sirtori, CR; Calabresi, L; Pisciotta, L; Cattin, L; Pauciullo, P; Montagnani, M; Manzato, E; Bittolo Bon, G; Fellin, R. Effect of statins on LDL particle size in patients with familial combined hyperlipidemia: a comparison between atorvastatin and pravastatin. *Nutr. Metab. Cardiovasc. Dis.,* 2005, 15, 47-55.

[110] van Tits, LJ; Smilde, TJ; van Wissen, S; de Graaf, J; Kastelein, JJ; Stalenhoef, AF. Effects of atorvastatin and simvastatin on low-density lipoprotein subfraction profile, low-density lipoprotein oxidizability, and antibodies to oxidized lowdensity lipoprotein in relation to carotid intima media thickness in familial hypercholesterolemia. *J. Investig. Med.,* 2004, 52, 177-184.

[111] Marz, W; Scharnagl, H; Abletshauser, C; Hoffmann, MM; Berg, A; Keul, J; Wieland, H; Baumstark, MW. Fluvastatin lowers atherogenic dense low-density lipoproteins in postmenopausal women with the atherogenic lipoprotein phenotype. *Circulation,* 2001, 103, 1942-1948.

[112] Bays, HE; McGovern, ME. Once-daily niacin extended release/lovastatin combination tablet has more favorable effects on lipoprotein particle size and subclass distribution than atorvastatin and simvastatin. *Prev. Cardiol.,* 2003, 6, 179-188.

[113] Rizzo, M; Vekic, J; Koulouris, S; Zeljkovic, A; Jelic-Ivanovic, Z; Spasojevic-Kalimanovska, V; Rini, GB; Sakellariou, D; Pastromas, S; Mikhailidis, DP; Manolis AS. Effects of rosiglitazone on fasting and postprandial low- and high-density lipoproteins size and subclasses in type 2 diabetes. *Angiology,* 2010, 61, 584-590.

[114] Rizzo, M; Corrado, E; Patti, AM; Rini, GB; Mikhailidis, DP. Cilostazol and atherogenic dyslipidemia: a clinically relevant effect? *Expert Opin. Pharmacother.,* 2011, 12: 647-655.

[115] Mosca, L. C-reactive protein - to screen or not to screen? *N. Eng. J. Med.,* 2002, 347, 1615-1617.

[116] Wilson, PWF; Myers, GL; Cooper, GR; Grundy, SM; Labarthe, DR. Lipoprotein subclasses and particle concentration and cardiovascular disease risk. In: Myers GL, editor. Laboratory medicine practice guidelines: emerging biomarkers for primary prevention of cardiovascular disease and stroke. Washington, DC: AACC press; 2009; 25-27.

[117] Kushner, FG; Hand, M; Smith, SC Jr; King, SB; Anderson, JL; Antman, EM; Bailey, SR; Bates, ER; Blankenship, JC; Casey, DE Jr; Green, LA; Hochman, JS; Jacobs, AK; Krumholz, HM; Morrison, DA; Ornato, JP; Pearle, DL; Peterson, ED; Sloan, MA; Whitlow, PL; Williams, DO. 2009 focused updates: ACC/AHA guidelines for the management of patients with ST elevation myocardial infarction (updating the 2004 guideline and 2007 focused update) and ACC/AHA/SCAI guidelines on percutaneous coronary intervention (updating the 2005 guideline and 2007 focused update) a report of the American College of Cardiology Foundation/American Heart Association Task Force on Practice Guidelines. *J. Am. Coll Cardiol.,* 2009, 54, 2205-2241.

[118] Myers, GL; Christenson, RH. Introduction. In: Myers GL, editor. Laboratory medicine practice guidelines: emerging biomarkers for primary prevention of cardiovascular disease and stroke. Washington, DC: AACC press; 2009; 1-3.

[119] Hirano, T; Ito, Y; Saegusa, H; Yoshino, G. A novel and simple method for quantification of small dense LDL. *J. Lipid. Res.*, 2003, 44, 2193-2201.

[120] Hirano, T; Ito, Y; Yoshino, G. Measurement of small, dense low-density lipoprotein particles. *J. Atheroscler. Thromb.,* 2005, 12, 67-72.

[121] Ogita, K; Ai, M; Tanaka, A; Ito, Y; Hirano, T; Yoshino, G; Shimokado, K. Circadian rhythm of serum concentration of small dense low-density lipoprotein cholesterol. *Clin. Chim. Acta*, 2007, 376, 96-100.

[122] Ogita, K; Ai, M; Tanaka, A; Ito, Y; Hirano, T; Yoshino, G; Shimokado, K. Serum concentration of small dense low-density lipoprotein-cholesterol during oral glucose tolerance test and oral fat tolerance test. *Clin. Chim. Acta,* 2008, 387, 36-41.

[123] Koba, S; Hirano, T; Ito, Y; Tsunoda, F; Yokota Y; Ban Y; Iso Y; Suzuki H; Katagiri T. Significance of small dense low-density lipoprotein-cholesterol concentrations in relation to the severity of coronary heart diseases. *Atherosclerosis,* 2006, 189, 206-214.

[124] Nozue, T; Michishita, I; Ishibashi, Y; Ito, S; Iwaki, T; Mizuguchi, I; Miura, M; Ito, Y; Hirano, T. Small dense low-density lipoprotein cholesterol is a useful marker of metabolic syndrome in patients with coronary artery disease. *J. Atheroscler. Thromb.,* 2007, 14, 202-207.

[125] Hirano, T; Nohtomi, K; Sato, Y; Kamata, K; Ito, Y. Small dense LDL-cholesterol determined by a simple precipitation assay for screening familial combined hyperlipidemia. *Atherosclerosis,* 2009, 205, 603-607.

[126] Shoji, T; Hatsuda, S; Tsuchikura, S; Shinohara, K; Kimoto, E; Koyama, H; Emoto, M; Nishizawa, Y. Small dense low-density lipoprotein cholesterol concentration and carotid atherosclerosis. *Atherosclerosis,* 2009, 202, 582-588.

[127] Tokuno, A; Hirano, T; Hayashi, T; Mori, Y; Yamamoto, T; Nagashima, M; Shiraishi, Y; Ito, Y; Adachi, M. The effects of statin and fibrate on lowering small dense LDL-cholesterol in hyperlipidemic patients with type 2 diabetes. *J. Atheroscler. Thromb.*, 2007, 14, 128-132.

In: Dyslipidemia ISBN: 978-1-62100-917-7
Editors: Z. Ačimovič et.al, pp. 175-197

Chapter 6

THE ROLE OF LIGAND-ACTIVATED TRANSCRIPTION FACTORS PPAR, LXR, CAR, AND PXR IN THE DEVELOPMENT OF METABOLIC SYNDROME

Elena N. Pivovarova
Institute of Cytology and Genetics,
Siberian Branch of Russian Academy of Sciences,
Novosibirsk, Russia

ABSTRACT

It is known that the metabolic syndrome (MS) leads to serious cardiovascular disease which continues to be the number one cause of death in developed countries. MS is characterized by dyslipidemia, hypertension, glucose tolerance, and obesity. The MS distribution is growing catastrophically, but molecular mechanisms responsible for developments of complex impairments in MS still remain basically poorly investigated. The formation of complex MS symptoms suggests systemic impairments in lipid and carbohydrate metabolism; it appears that these impairments should have a common basis at the level of expression of appropriate genes. Expression of genes involved into lipid and carbohydrate metabolism is regulated by various transcription factors, including those referred to the superfamily of nuclear hormone receptors: peroxisome proliferator-activated receptors (PPAR), liver X receptors (LXR), pregnane X receptors (PXR), and constitutive androstane receptors (CAR). PXR and CAR are mainly known as the sensors of xenobiotics. Since these transcription factors are ligand-activated, they represent perspective targets for pharmacological treatments. In the recent time, the search of natural and synthetic PPAR ligands is in intensive process. There is a cross-talk between signal transductionpathways of PPAR, LXR, PXR, and CAR; this suggests their integrated role in regulation of genes of lipid and carbohydrate metabolism. In our work at the model of rat strain with Inherited Stress-Induced Arterial Hypertension which demonstrates all signs of MS development it was shown that an increased content of triglyceride, VLDL and LDL cholesterol, a decreased content of HDL cholesterol, a high level of apoB-100, and a decreased level of apoA-I, increased body weight and blood glucose level are associated with increased DNA-binding activity of PPAR, LXR, PXR, and CAR in the liver; this suggests involvement of these transcription factors in the development of MS in ISIAH rats. We have also shown that stress can affects the DNA-binding activity of these transcription factors. In this review the role of ligand-activated transcription factors PPAR, LXR, CAR, PXR and the role of stress in the development of metabolic syndrome will be discussed.

INTRODUCTION

The metabolic syndrome (MS) is a complex of pathologies that increase the risk of cardiovascular diseases which still are the leading cause of mortality in industrial countries. It is known that MS is characterized by dyslipidemia, hypertension, glucose tolerance, and obesity.

In 1980th, Reaven [Reaven, 1988] proposed to pool such metabolic impairments under a common name of the X syndrome, which is now known as the metabolic syndrome.

Although the term "metabolic syndrome" is used for several years, only recently attempts to give clear definition of this term have been undertaken. According to the National Cholesterol Education Program-Adult Treatment Panel III (NCEP-ATP III) criteria of the metabolic syndrome include three or more impairments including visceral obesity, increase in blood triglycerides, decrease in blood HDL cholesterol, hypertension, and increased blood glucose (NCEP-ATP III, 2001).

NCEP ATP III has defined the criteria for MS in order to assist physicians in diagnosing patients at risk for coronary heart disease and diabetes mellitus type 2, are clustered metabolic risk factors (Grundy, 2005). NCEP ATP III identified five easily measurable criteria for identifying patients with MS: the levels of triglycerides and fasting plasma glucose, HDL cholesterol, blood pressure and waist circumference (Table 1).

The prevalence of MS in the U. S. based on the data from National Health and Nutrition Examination Survey (NHANES) collected between 1999 and 2002 was 34.5% of population sample (n = 3601) of men and women over age 20. The prevalence increases with the age. Among men and women over 50 years MS distribution (according data 1999-2002) was approximately 50% (Ford, 2005).

Increasing prevalence of MS in most cases there is due to an increase in body weight. It is likely that the increase in body weight can change each of the metabolic risk factors, but there is evidence that the severity of each risk factor is under its own genetic control.

Some researchers believe that insulin resistance is the only major factor responsible for the metabolic syndrome (Reaven, 2004). However, regardless of root cause diagnosis of metabolic syndrome indicates an increased risk of cardiovascular disease. The magnitude of risk varies depending on the presence of risk factors and their severity (Grundy, 2005).

Insulin resistance is common among people with high triglycerides and low HDL cholesterol - two the most common risk factors among people diagnosed with MS (Ford, 2002). However, fasting hyperglycemia occurs only in 12% of people that meet diagnostic criteria for MS, because this is the last of the risk factors developing in patients with MS. A simple indicator of insulin resistance in people with MS is the ratio of fasting triglyceride concentration of HDL cholesterol. The value of triglyceride/HDL cholesterol = 3. 5 points to insulin resistance as authentically as an increased level of fasting insulin (McLaughlin, 2005).

Another risk factor for MS is abdominal obesity (Grundy, 2005). Abdominal obesity contributes to increased insulin resistance, resulting in increased flux of free fatty acids from adipose tissue to the liver and insulin resistance in the muscles.

MS is associated with an increased relative risk of subsequent cardiovascular diseases by 65% and increases the risk of development of type 2 diabetes mellitus three times (Ford, 2005a).

The MS distribution is growing, but molecular mechanisms responsible for developments of complex impairments in MS still remain basically poorly investigated.

The development of complex MS symptoms suggests systemic impairments in lipid and carbohydrate metabolism; it appears that these impairments should have a common basis and it have to be at the level of expression of appropriate genes.

Expression of genes involved into lipid and carbohydrate metabolism is regulated by the number of transcription factors, including those referred to the superfamily of nuclear hormone receptors: peroxisome proliferator-activated receptors (PPAR) (Michalik, 2006;Lefebvre , 2006), liver X receptors (LXR) [5], pregnane X receptors (PXR), constitutive androstane receptors (CAR) [6].

Nuclear hormone receptors are ligand-activated transcription factors that regulate gene expression by interacting with specific DNA sequences upstream of their target genes.This sequence was termed the response element (RE). Nuclear hormone receptors regulate gene expression in response to a wide range of developmental, physiological, and environmental cues.

The PPAR, LXR, PXR, and CAR are activated after their interaction with corresponding ligands. PPAR, LXR, and CAR are localized in cytoplasm and after interaction with their ligands are translocated into the nucleus. PXR is localized in the nucleus where it is activated by its ligand. In the nucleus, after heterodimer formation with RXR protein (retinoid X receptor) nuclear receptors are specifically bound to regulatory sites of target genes and modulate their expression (Michalik, 2006). Expression of transcription factors and their iso-forms is tissue- and organ-specific.

Peroxisome Proliferator-Activated Receptor (PPAR)

PPAR has three isoforms: PPAR-alpha, PPAR-beta, and PPAR-gamma. PPAR-alpha is expressed predominantly in the liver, kidney and heart, and is primarily involved in fatty acid oxidation. PPAR-beta (delta) is abundantly and ubiquitously expressed. PPAR-gamma is most abundantly expressed in adipose tissue, with less expression in the colon and immune system. As well as in adipocytes and T cells, PPAR-gamma is also expressed in endothelial cells, vascular smooth muscle cells, and macrophages. Small quantities of PPAR-gamma1 are expressed in the liver; PPAR-gamma2 is expressed only in adipose tissue (Michalik, 2006).

PPARs were found to be mediators of pharmacologic agents that induce hepatocyte peroxisome proliferation. Peroxisomes are organelles that participate in fatty acid metabolism. Clofibrate analogues, hypolipidemicagents that control plasma cholesterol and triglyceride levels, can induce proliferation of liver cell peroxisomes (Reddy, 1973; Reddy, 1974).

In addition, two lipid-lowering compounds structurally different from clofibrate, [4-chloro-6-(2,3-xylidino)-2-pyrimidinylthio]acetic acid (Wy-14,643) and 2-chloro-5-(3,5-dimethylpiperidino-sulfonyl)benzoic acid (tibric acid), also were found to stimulate hepatocyte peroxisome proliferation (Reddy, 1975). Although hypolipidemic drugs were demonstrated to activate peroxisome proliferation, these studies did not establish a mechanism. Subsequent studies identified a protein whereby peroxisome proliferators bind with affinity (Lalwani, 1987), and this protein was later identified as a member of the nuclear hormone-receptor superfamily that includes steroid, retinoid, and thyroid hormone receptors (Evans, 1988). The name peroxisome proliferator-activated receptor took origin from the cloning by Issemann*et al.* (Issemann, 1990) to identify possible endogenous mediators of peroxisome proliferation–induced gene transcription in rodent livers.

PPAR-alpha regulates liver and skeletal muscle lipid metabolism as well as glucose homeostasis. Acting as a molecular sensor of endogenous fatty acids (FAs) and their derivatives, this transcription factor regulates the expression of genes encoding enzymes and transport proteins controlling lipid homeostasis, thereby stimulating FA oxidation and improving lipoprotein metabolism. PPAR-alpha also exerts pleiotropic antiinflammatory and antiproliferative effects and prevents the proatherogenic effects of cholesterol accumulation in macrophages by stimulating cholesterol efflux (Lefebvre, 2006).

PPAR- alpha controls expression of genes involved in intra- and extracellular lipid metabolism including carnitinepalmitoyltransferase 1responsible for transport of fatty acids into mitochondria; the class B scavenger receptor CD36 involved into high-affinity peripheral fatty acid uptake; fatty acid transport protein; acyl-CoA synthetase converting fatty acids into their acyl-CoA derivatives; lipoprotein lipase catalyzing cleavage of chylomicron and VLDL triglycerides; apoA-I and apoA-II, the main HDL protein, thus influencing HDL cholesterol level (Lefebvre, 2006). PPAR-alpha also controls expression of genes encoding enzymes involved into glycerol conversion into glucose in the liver: glycerol 3-phosphate dehydrogenase and glycerol kinase and also genes encoding glycerol, as well as aquaporin 3 and 9 transporters (Patsouris, 2004).

PPAR-beta regulates expression of genes involved into fatty acid catabolism (Michalik, 2006).

Table 1. The criteria for clinical diagnosis of metabolic syndrome

Indices	Limiting values
Waist circumference	Men: ≥ 102 cm (≥ 40 inches) Women: ≥ 88 cm (≥ 35 inches)
Increased triglyceride levels	≥ 1.7 mM / l (≥ 150 mg/dl)
Reduced HDL cholesterol	Men: < 1.03 mmol/l (< 40 mg/dl) Women: < 1.3 mmol/l (< 50 mg/dl)
High blood pressure	Systolic blood pressure ≥ 130 mm Hg or diastolic blood pressure ≥ 85 mm Hg
Elevated fasting glucose	≥ 5.6 mmol/l (≥ 100 mg/dl)

PPAR-gamma participates in expression of genes involved in lipid accumulation and also glucose metabolism (Michalik, 2006). Although PPAR-gamma was initially found to be critical for adipocyte differentiation and function, over time, PPARγ was discovered to play an important role in the cardiovascular system (Hamblin, 2009). PPAR-gamma is involved in the regulation of lipid metabolism, as ligand-dependent activation leads to an increase in genes that regulate fatty acid uptake and storage (Semple, 2005). PPAR-gamma plays a role in glucose homeostasis and insulin sensitivity (Forman, 1995). PPAR-gamma is mainly associated with adipose tissue, where it controls adipocyte differentiation and insulin sensitivity. It has been shown that PPAR-gamma can facilitate differentiation of fibroblasts into adipocytes (Chawla, 1994).

PPAR-alpha overexpression in the heart results in insulin resistance in heart and liver; its overexpression in muscles leads to glucose tolerance in muscles. PPAR-gamma overexpression in hepatocytes causes hepatic steatosis (fatty infiltration) due to triglyceride accumulation (Michalik, 2006). Hepatic steatosis is often associated with MS (Reddy, 2006).

It was shown that expression of PPAR- alpha and PPAR-gamma is increased in the vessels of hypertensive SHR (Spontaneously Hypertensive Rats) strain compared with the control WKY (Wistar-Kyoto) strain (Diep, 2001).

The endogenous ligands of PPAR are saturated fatty acids, native and oxidized unsaturated fatty acids, native and oxidized eicosanoids, prostaglandins, and prostacyclins.

Activators of PPAR-alpha (fibrates) and -gamma (thiazolidinediones) have been used clinically for a number of years in the treatment of hyperlipidaemia and to improve insulin sensitivity in diabetes. More recently,

PPAR activation has been found to confer additional benefits on endothelial function, inflammation and thrombosis, suggesting that PPAR agonists may be good candidates for the treatment of cardiovascular disease. In this regard, it has been demonstrated that PPAR activators are capable of reducing blood pressure and attenuating the development of atherosclerosis and cardiac hypertrophy (Robinson, 2009).

Cellular and animal models of PPAR-alpha help explain the clinical actions of fibrates, synthetic PPAR-alpha agonists used to treat dyslipidemia and reduce cardiovascular disease and its complications in patients with the metabolic syndrome. Although these preclinical studies cannot predict all of the effects of PPAR-alpha in humans, recent findings have revealed potential adverse effects of PPAR-alphaaction, underlining the need for further study (Lefebvre, 2006).

LIVER X RECEPTOR (LXR)

There are two isoforms of LXR: LXR-alpha and LXR-beta. LXR-alpha is expressed most highly in the liver and to a lesser extent in the kidney, small intestine, spleen, and adrenal gland (Willy, 1995). In contrast to the restricted expression pattern of LXR-alpha, LXR-betais ubiquitously expressed (Kainu, 1996). These two isoforms of LXR act as the regulators of lipid and glucose metabolism. LXRs have recently been suggested to play a novel role in the regulation of drug metabolism. CAR forms regulatory network with LXR in exerting its effect on lipid metabolism (Xiao, 2010).

LXRs have been shown to regulate the expression of a battery of metabolic genes, especially those involved in lipid metabolism (Xiao, 2010). LXRs function as cholesterol sensors, which protect mammals from cholesterol overload.

In the liver, excess of cholesterol is converted into bile acids and is removed from the cells, while the synthesis of cholesterol and lipoprotein capture decrease. Functional binding sites for LXR-alpha and LXR-beta were found in promoters of a number of genes encoding key enzymes, transporters, and regulators of these processes. LXR regulates expression of SREBP-1, the protein controlling expression of fatty acid synthase and other key genes involved into fatty acid biosynthesis (Herzog, 2007). Moreover, LXR may directly regulate expression of fatty acid synthase (Tobin, 2000). LXR also regulates expression of the gene encoding ABCA-1, the protein involved into apoA-I-mediated cholesterol transport from hepatocytes (Chawla, 2001).

It was shown that LXR is involved into regulation of glucose metabolism by modulating expression of genes encoding key enzymes of gluconeogenesis (phosphoenolpyruvatecarboxykinase and glucose-6-phosphatase) regulating blood glucose level (Cao, 2003).

The endogenous ligands of LXR are oxidized cholesterol derivatives. LXR regulate the nutrient metabolism pathways through their interactions with specific, naturally occurring oxysterols, including 22(*R*)-hydroxycholesterol, 24(*S*)-hydroxycholesterol and 24(*S*),25-epoxycholesterol (Lehmann, 1996; Peet, 1998).

LXRs were shown to function as sterol sensors protecting the cells from cholesterol overload by stimulating reverse cholesterol transport and activating its conversion to bile acids in the liver.

This finding has led to identification of LXR agonists as potent antiatherogenic agents in rodent models of atherosclerosis. However, first-generation LXR activators were also shown to stimulate lipogenesis via sterol regulatory element binding protein-1c leading to liver steatosis and hypertriglyceridemia. Despite their lipogenic action, LXR agonists possess antidiabetic properties. LXR activation normalizes glycemia and improves insulin sensitivity in rodent models of type 2 diabetes and insulin resistance. Antidiabetic action of LXR agonists is thought to result predominantly from suppression of hepatic gluconeogenesis. However, recent studies suggest that LXR activation may also enhance peripheral glucose uptake (Baranowski, 2008).

As regulators of metabolism, LXRs have been considered as potential drug targets by the pharmaceutical industry, and synthetic LXR ligands have been developed, which are widely used as tools in biomedical research. Synthetic LXR ligands include T0901317 and GW3965 (Baranowski, 2008).

Pregnane X Receptor (PXR) and Constitutive Androstane Receptor (CAR)

PXR and CAR are two closely related and liver-enriched nuclear hormone receptors. PXR and CAR are expressed predominantly in the liver and intestines where their target genes are also located (di Masi, 2009).

PXR and CAR are mainly known as the sensors of xenobiotics and play a key role in xenobiotic detection and response. The xenobiotic response represents a complex group of chemical reactions aimed to inactivate and eliminate foreign chemicals. Xenobiotic response is also active against endogenous products of metabolism.

Recent results suggest that PXR and CAR also have important endobiotic roles in energy metabolism by affecting the metabolism of fatty acids, lipids and glucose. PXR and CAR exert their effects on energy metabolism through direct gene regulation or through crosstalk with other transcriptional regulators (Wada, 2009).

CAR and PXR modulate beta-oxidation of fatty acids, PXR controls expression of transcription factors and enzymes involved into lipogenesis (Moreau, 2008) and also influence on transcription of the scavenger receptor CD36 gene (Zhou, 2006). At the same time CAR and PXR influence on transcription factors and cofactors, involved into regulation of genes encoding enzymes of gluconeogenesis in the liver (Moreau, 2008).

CAR and PXR are transcription factors activated by a variety of endogenous and exogenous ligands. Both CAR and PXR elevate the expression of the detoxification machinery in the presence of endogenous and exogenous ligands, regulating the expression of Phase I and II metabolizing enzymes and of phase III transporters. Prescription drugs are included in the list of xenobiotic recognized by CAR and PXR. Among others, statins represent a good example of CAR- and PXR-mediated drug metabolism. Indeed statins, the most effective and prescribed cholesterol-lowering drugs, are target of xenobiotic response mediated by CAR and PXR (Marino, 2011).

The PXR has also recently been shown to play an endobiotic role by impacting lipid homeostasis (Zhou, 2006;Zhou, 2008). Expression of an activated PXR in the livers of transgenic mice resulted in an increased hepatic deposit of triglycerides. This PXR-mediated lipid accumulation was independent of the activation of the lipogenic transcriptional factor sterol regulatory element-binding protein 1c (SREBP-1c) and its primary lipogenic target enzymes, including fatty acid synthase (FAS) and acetyl CoA carboxylase 1 (ACC-1). Instead, the lipid accumulation in transgenic mice was associated with an increased expression of the free fatty acid transporter CD36 and several accessory lipogenic enzymes, such as stearoyl CoA desaturase-1 (SCD-1) and long-chain free fatty acid elongase (FAE).

The activation of PXR was also associated with an inhibition of pro-β-oxidative genes, such as peroxisome proliferator-activated receptor α (PPARα) and thiolase, and an upregulation of PPARγ, a positive regulator of CD36. The crossregulation of CD36 by PXR and PPARγ suggests that this fatty acid transporter may function as a common target of orphan nuclear receptors in their regulation of lipid homeostasis (Zhou, 2008).

Unlike other nuclear receptors, such as the steroid receptors (e.g., estrogen receptors-α and -β), which interact selectively with their physiological ligands, PXR ligands are structurally diverse and include prescription drugs, herbal medicines, dietary supplements, environmental pollutants and endobiotics (Kliewer, 2002).

Many PXR ligands have been identified among prescription drugs, and include the antibiotics rifampicin (RIF), clotrimazole and ritonavir; the antineoplastic drugs cyclophosphamide, taxol and tamoxifen; the endocrine drugs cyproterone acetate and RU486; the anti-inflammatory agent dexamethasone; the anti-type 2 diabetes drug troglitazone; the antihypertensive drugs nifedipine and spironolactone; and the sedatives glutethimide and phenobarbital (Willson, 2002). Among dietary supplements, vitamins K2 and E have been established as weak PXR activators (Tabb, 2003;Landes, 2003). Several groups also reported that a number of environmental pollutants are PXR ligands, such as organochlorine pesticides and polybrominateddiphenyl ether fame retardants (Coumoul, 2002). In addition, some endobiotics were identified as PXR ligands, including certain bile acids, bile acid precursors and estrogens (Mnif, 2007;Goodwin, 2003).

CARs play roles in regulating expression of CYP genes in response to exogenous xenobiotics and endogenous lipid compounds. There is overlap between CAR and PXR target genes (Maglich, 2002). For example, PXR regulates the expression of both *CYP2B6* and *CYP3A4*, whereas CAR preferentially regulates *CYP2B6*as a consequence of its weaker binding to the PXR response element in the *CYP3A4* promoter (Faucette, 2006). Other examples of CAR-regulated genes include *CYP2C8, CYP2C9*, and *CYP2C19,* phase II conjugation enzymes, such as UDP-glucuronosyltransferase*UGT1A1*, sulfotransferase*Sult2a1*, and glutathione *S*-transferases*Gsta1*(Maglich, 2002). CAR has also been shown to regulate the repression of enzymes involved in gluconeogenesis, such as phosphoenoylpyuvatecarboxykinase 1 (PEPCK1), and beta-oxidation enzymes, such as carnitinepalmitoyltransferase 1 (Tien, 2006). In contrast to PXR, CAR is constitutively active (Timsit, 2007).

CAR can repress the genes that encode enzymes involved in signal transduction, fatty acid oxidation, and/or energy metabolism, *e.g.* phosphoenolpyruvatecarboxykinase 1, enoyl coenzyme A isomerase, and carnitinepalmitoyltransferase 1 (*Ueda, 2002).*

Original ligands identified for CAR include the testosterone metabolites 5α-androstan-3α-ol (androstanol) and 5α-androst-16-en-3α-ol (androstenol) (Forman, 1998). Unlike classical NRs that are ligand activated or deactivated, CAR activity could be affected by both ligand-dependent and -independent pathways (Li, 2010). CAR has become best known for its ability to regulate induction of the *CYP2B* gene family by phenobarbital (PB) and PB-like inducers [*e.g.* chlorpromazine, phenytoin, dichlorodiphenyltrichloroethane, 1,4-bis [2-(3,5-dichloropyridyloxy)]benzene (TCPOBOP) and polychlorinated biphenyls], thereby revealing its function as a xenobiotic-sensing nuclear receptor (*Sueyoshi, 1999)*.

CAR has the potential to impact numerous signaling pathways via the genes it modulates directly and by its interference with other nuclear receptor signaling pathways. This creates a unique integrative mechanism to modulate the metabolism of not only xenobiotic substances but also endogenously produced steroids and dietary factors. Unlike its closest relative PXR, the function of which relies solely on ligand binding, CAR functions ligand independently and can be regulated by both direct ligand binding and indirect activation processes (Swales, 2004).

MS Modeling

A high percentage of MS morbidity determines much attention to MS modeling, mechanisms of MS development, and new approaches to MS treatment. Although, several inbred, congenic and transgenic rat strains modeling human MS are known, none of existing experimental models gives complete description of mechanisms underlying disease complexity in humans.

It is known that chronic stress causes systemic disturbances in lipid metabolism and cardiovascular functions and plays an important role in the development of MS in humans Chandola, 2006). A rat strain with inherited stress induced arterial hypertension (ISIAH rats) was obtained at the Institute of Cytology and Genetics (Siberian Branch of Russian Academy of Sciences). The ISIAH rat strain was selected from Wistar rats by the level of systolic arterial blood pressure under conditions of mild emotional stress. In ISIAH and Wistar rats basal levels of blood pressure are 150 - 160 and 120 mm Hg, respectively. Mild emotional stress, immobilization in a narrow cage for 30 min, causes a significant increase of blood pressure in ISIAH rats up to 200 mm Hg, whereas in WAG rats such stress insignificantly influences this parameter (Fedoseeva, 2011).

In our work it was shown that ISAH rat strain demonstrates all signs of MS development (Pivovarova, 2011). Hypertensive ISIAH rats have statistically lower HDL cholesterol and increased triglyceride level than the normotensive WAG rats (by 16.2 and 34.7%, respectively; Table 1). Such changes in the lipid spectrum are typical for MS and are atherogenic signs [19].

Analysis of cholesterol distribution in lipoprotein fractions (Table 2) also revealed the atherogenic tendency in ISIAH rats. These animals were characterized by increased cholesterol content in the athero-genic lipoprotein fractions (VLDL and LDL, by 46 and 31%, respectively) and decreased cholesterol content in the anti-atherogenic lipoprotein fractions: HDL_2 and HDL_3 (by 33 and 18%, respectively).

The level of apoA-I, the main HDL protein, was lower in blood of ISIAH rats (by 24%) as compared with WAG rats. The level of apoB-100, the main LDL protein, was higher (by 49.5%) in blood of ISIAH rats than in WAG rats (Table 2). The decreased level of apoA-I and the increased level of apoB-100 are also risk factors for development of cardiovascular diseases [19].

An average body mass of ISIAH rats was higher (by 8.5%) than that of age matched WAG rats. A basal glucose level in ISIAH rats was also higher (by 15%) than that of WAG rats (Table 3). Increased body weight and blood glucose level are characteristic signs of MS (NCEP-ATP, 2001).

According to literature data, a hypertensive strain of rats known as SHR (Spontaneously Hypertensive Rats) is characterized by increased LDL cholesterol content compared with the normotensive WKY (Wistar-Kyoto) strain (Yu, 1993). Another hypertensive rat strain known as Lyon hypertensive differs from the normotensive control (Lyon normotensive rats) by increased body weight and increased blood triglyceride content (Sassolas, 1981). Insulin resistance was demonstrated in several hypertensive animal models including SHR (Shimamoto, 2006).

Table 2. Comparative study of hypertensive ISIAH and normotensive WAG rats: lipid spectrum, body mass, and glucose level

Indices	ISIAH compared to WAG, %
Total cholesterol	no significant differences
HDL cholesterol	decreased by 16.2%
Triglycerides	increased by 34.7%
Body mass	increased by 18.5%
Basal glucose level	increased by 15%

Table 3. Comparative study of hypertensive ISIAH and normotensive WAG rats: content of cholesterol in separate lipoprotein fractions and relative content of apoA-I and apoB-100

Indices		ISIAH compared to WAG, %
Atherogenic lipoprotein fractions	LDL	increased by 46%
	DL	increased by 31%
	ApoB-100	increased by 49.5%
Anti-atherogenic lipoprotein fractions	HDL_2	decreased by 33%
	HDL_3	decreased by 18%
	ApoA-I	decreased by 24%

Table 4. Comparative study of hypertensive ISIAH and normotensive WAG rats: functional activity of transcription factors in the liver nuclear extracts

Transcription factor	DNA-binding activity, ISIAH compared to WAG
PPAR	increased by 9.4-fold
LXR	increased by 3.9-fold
CAR	increased by 7.1-fold
PXR	increased by 4.8-fold

Thus, rats obtained during selection for increased blood pressure have some symptoms typical for MS. ISIAH rats used in this study differ from all other hypertensive strains: this strain was selected for enhanced response of blood pressure to the emotional stress. More pronounced MS symptoms observed in ISIAH rats suggest an essential role of the emotional stress in MS development (Pivovarova, 2011).

Table 5. The DNA-binding activity of the transcription factors in the liver nuclear extracts of hypertensive ISIAH and normotensive WAG rats after stress

Transcription factor	DNA-binding activity	
	ISIAH	WAG
PPAR	not changed	increased
LXR	not changed	increased
CAR	decreased	increased
PXR	decreased	increased

Possible Mechanisms of MS Development and Approaches to the MS Treatment

The MS distribution is growing catastrophically, but mechanisms responsible for developments of complex impairments in MS still remain basically poorly investigated.

The formation of complex MS symptoms suggests systemic impairments in lipid and carbohydrate metabolism; it appears that these impairments should have a common basis at the level of expression of appropriate genes.

It was shown that ligand-activated transcription factors PPAR, LXR, PXR, and CAR participate in regulation of expression of genes involved into lipid and carbohydrate metabolism (Lefebvre, 2006; Herzog, 2007; Moreau, 2008;Van der Leij, 2007). All of this transcription factors are expressed in the liver.

The liver is the main organ controlling energy homeostasis of the body. The liver maintains triglyceride homeostasis by storing circulating fatty acids as triglycerides, oxidizing fatty acids and secreting triglycerides as VLDL. Under normal conditions, the major sources of energy to maintain the whole-body energy homeostasis are triglycerides converted from fatty acids and glucose produced in the liver. During starvation or exercise, fatty acids are released from adipocytes and glucose is produced in the liver to be used by skeletal muscle to meet the body's energy requirements. Hepatic lipid metabolism involves the synthesis of lipids (lipogenesis), fatty acid oxidation (b-oxidation) and lipid secretion. Lipogenesis is defined as de novo fat synthesis and includes fatty acid synthesis and subsequent conversion of fatty acids to triglycerides in the liver and adipose tissue. Gluconeogenesis provides another major source of energy. Hepatic gluconeogenesis is tightly linked to fasting and starvation. Disruption of lipid and glucose metabolism in the liver might trigger various cardiovascular and metabolic diseases, such as atherosclerosis, type II diabetes, obesity and insulin resistance (Wada, 2009).

In our work at the model of rat strain with Inherited Stress-Induced Arterial Hypertension (ISIAH) which demonstrates all signs of MS functional activity of PPAR, LXR, PXR, and CAR in the liver has been evaluated by their ability to bind oligonucleotides, which imitate binding sites (response elements) in the regulatory regions of target genes (Pivovarova, 2011). The DNA-binding activity (functional activity) of transcription factors was determined by Electrophoretic Mobility Shift Assay(EMSA). Nuclear protein extracts were isolated from liver cells. Nuclear extracts were incubated with labeled oligonucleotides (600-1000 cpm per sample). Native electrophoresis was performed in 4.5% PAAG. After electrophoresis gel was fixed, dried, and radioautographed.

The DNA binding activity of PPAR, LXR, PXR, and CAR was higher in liver nuclear extracts of ISIAH rats than in WAG rats. Quantitative densitometry of radioautographs has shown that in ISIAH rats the binding activity of PPAR, LXR, PXR, and CAR was higher by 9.4-, 3.9, 4.8, and 7.1-fold correspondingly as compared to WAG rats (Table 4).

Thus, observed signs of MS in ISIAH rats are associated with increased DNA-binding activity of PPAR, LXR, PXR, and CAR; this suggests involvement of these transcription factors in the development of MS in ISIAH rats (Pivovarova, 2011).

Since the stress plays a crucial role in MS development (Chandola T. et all, 2006), we investigated the influence of mild emotional stress on functional activity of PPAR, LXR, PXR, CAR in the liver.

Male rats (4-5 months old) of stress-sensitive hypertensive ISIAH and normotensive WAG (Wistar Albino Glaxo) strains were used in this study. The rats of both strains were divided into two parts (four groups). One part was put into stress by restriction of animals in narrow cylindrical cages. Another part was a reference intact group.

The DNA-binding activity of PPAR and LXR was increased in WAG rats after stress, whereas DNA-binding activity of these transcription factors did not change in ISIAH rats after stress. The DNA-binding activity of CAR and PXR was increased in WAG rats after stress, whereas DNA-binding activity of these transcription factors was decreased in ISIAH rats after stress (Table 5).

Our data suggests that the stress can influence DNA-binding activity of transcription factors PPAR, LXR, PXR, and CAR which regulate expression of genes involved into lipid and carbohydrate metabolism.

These transcription factors are ligand-activated; there are a number of endogenous and exogenous ligands. Ligands for the transcription factors can be formed in the process of lipid metabolism. Lipoprotein lipase can hydrolyze lipoproteins, generating ligands for PPAR (Chawla, 2003). Oxidized LDL (ox-LDL) also can be the source of the ligands. It was shown that there is a correlation between the level of ox-LDL and the development of MS as a whole and its components such as abdominal obesity, hyperglycemia, and hypertriglyceridemia (Holvoet, 2008). It was also shown that ox-LDL can activate PPAR-alpha and PPAR-gamma (Taketa, 2008). Oxidized LDL trapped in the liver through scavenger receptor, where they are degraded (Mathieu, 2006); their oxidized lipid components can be ligands for transcription factors, affecting thus the network of genes involved in lipid and carbohydrate metabolism.

There is a cross-talk between signal transduction pathways of PPAR, LXR, PXR, and CAR (Woods, 2007); this suggests their integrated role in regulation of genes of lipid and carbohydrate metabolism.

In situation of an uncontrolled medication, as well as pollution of the environment the study of the role of transcription factors in the development of MS is especially important.

Thus, stress, nutrition, medicines, environment pollution via ligand-activated transcription factors can lead to the abnormalities in lipid and carbohydrate metabolism and MS development (Figure 1).

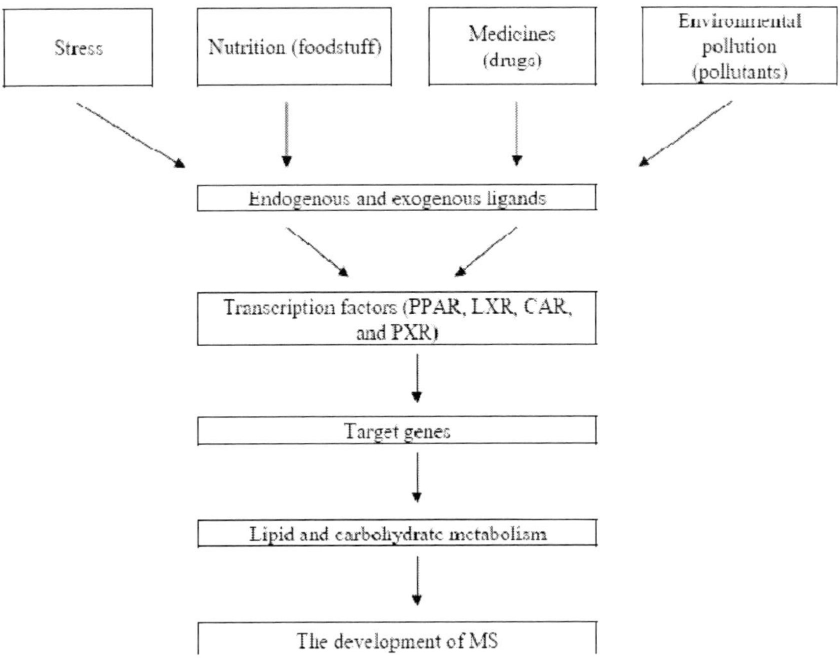

Figure 1. The role of transcription factors PPAR, LXR, CAR, and PXR in MS development.

Since PPAR, LXR, PXR, and CAR are ligand-activated receptors, they represent perspective targets for pharmacological treatments. At present, the search of natural and synthetic PPAR ligands is intensively performed (Michalik, 2006). Involvement of not only PPAR, but also LXR, PXR, and

CAR in the development of MS suggests that their signal transduction pathways should be taken into consideration during design of pharmacological preparations.

Conclusion

The development of complex MS symptoms suggests systemic impairments in lipid and carbohydrate metabolism. It appears that these impairments should have a common basis and it have to be at the level of expression of appropriate genes. Expression of genes involved into lipid and carbohydrate metabolism is regulated by ligand-activated transcription factors including those referred to the superfamily of nuclear hormone receptors: PPAR, LXR, PXR, and CAR. Since these transcription factors are ligand-activated, they represent perspective targets for pharmacological treatments. Complex studies of regulatory mechanisms, signaling pathways, and transcription targets of PPAR, LXR, PXR, and CAR can significantly help better understanding of mechanisms of MS development and provide valuable information for development of appropriate pharmacological approaches for MS therapy.

References

Baranowski M. Biological role of liver X receptors. *J. Physiol. Pharmacol.* 2008;59 Suppl 7:31-55.

Cao, G., Liang, Y., Broderick, C.L., Oldham, B.A., Beyer, T.P., Schmidt, R.J., Zhang, Y., Stayrook, K.R., Suen, C., Otto, K.A., Miller, A.R., Dai, J., Foxworthy, P. , Gao, H., Ryan, T.P., Jiang, X.C., Burris, T.P., Eacho, P.I., and Etgen, G.J., Antidiabetic action of a liver x receptor agonist mediated by inhibition of hepatic gluconeogenesis. *J. Biol. Chem.*, 2003, vol. 278, pp. 1131–1136.

Chandola, T. , Brunner, E., and Marmot, M., Chronic stress at work and the metabolic syndrome: prospective study.*B.M.J.*, 2006, vol. 332, pp. 521–525.

Chawla A. Schwarz EJ. Dimaculangan DD. Lazar MA. Peroxisome proliferator-activated receptor (PPAR) gamma: adipose-predominant expression and induction early in adipocyte differentiation. *Endocrinology*. 1994;135:798–800.

Chawla, A., Boisvert, W.A., Lee, C.H., Laffitte, B.A., Barak, Y., Joseph, S.B., Liao, D., Nagy, L., Edwards, P.A., Curtiss, L.K., Evans, R.M., and Tontonoz, P. A PPAR gamma-LXR-ABCA1 pathway in macrophages is involved in cholesterol efflux and atherogenesis. *Mol. Cell*, 2001, vol. 7, pp. 161–171.

Chawla A, Lee CH, Barak Y, He W, Rosenfeld J, Liao D, Han J, Kang H, Evans RM. PPARdelta is a very low-density lipoprotein sensor in macrophages. *Proc. Natl. Acad. Sci. USA*2003;100(3):1268-73.

CollinsJL, FivushAM, WatsonMA, GalardiCM, LewisMC, MooreLB, ParksDJ, WilsonG, TippinTK, BinzJG et al.Identification of a nonsteroidal liver X receptor agonist through parallel array synthesis of tertiary amines. *Journal of Medicinal Chemistry*2002,45:1963–1966.

Goodwin B, Gauthier KC, Umetani M, et al. Identification of bile acid precursors as endogenous ligands for the nuclear xenobiotic pregnane X receptor. *Proc. Natl. Acad. Sci. USA*2003;100:223–228.

Coumoul X, Diry M, Barouki R. PXR-dependent induction of human *CYP3A4* gene expression by organochlorine pesticides. *Biochem. Pharmacol.* 2002;64:1513–1519.

Diep, Q.N. and Schiffrin, E.L., Increased expression of peroxisome proliferator-activated receptor-alpha and -gamma in blood vessels of spontaneously hypertensive rats. *Hypertension*, 2001, vol. 38, pp. 249–254.

diMasi A, Marinis ED, Ascenzi P, et al. Nuclear receptors CAR and PXR: Molecular, functional, and biomedical aspects. *Mol. Aspects Med.* 2009;30(5):297–343.

Evans RM. The steroid and thyroid hormone receptor superfamily. *Science* (New York) 1988;240:889–895.

Faucette SR, Sueyoshi T, Smith CM, Negishi M, LeCluyse EL, Wang H. Differential regulation of hepatic *CYP2B6*and *CYP3A4* genes by constitutive androstane receptor but not pregnane X receptor. *J. Pharmacol. Exp. Ther.* 2006;317:1200–1209.

Fedoseeva, L. A., Ryazanova, M. A., Antonov, E. V., Dymshits, G. M. and Markel, A. L. Expression of the renin angiotensin system genes in the kidney and heart of ISIAH hypertensive rats. 2011. *Biochemistry (Moscow) Supplemental Series B: Biomedical Chemistry*Volume 5, Number 1, 37-43.

Ford E.S., Giles W.D., Dietz W.H. Prevalence of the metabolic syndrome among US adults. Findings from the Third National Health and Nutrition Examination Survey// *JAMA* - 2002.- Vol.287.- P.356-359.

Ford E.S. Prevalence of the metabolic syndrome defined by the International Diabetes Federation among adults in the U.S. *Diabetes Care* - 2005.- Vol.28.- P.2745-2749.

Ford E.S. Risks for all-cause mortality, cardiovascular disease, and diabetes associated with the metabolic syndrome. A summary of the evidence. *Diabetes Care* - 2005.- Vol.28.- P.1769-1778.

Forman BM. Tontonoz P. Chen J. Brun RP. Spiegelman BM. Evans RM. 15-Deoxy-delta 12, 14-prostaglandin J2 is a ligand for the adipocyte determination factor PPAR gamma. *Cell*1995;83:803–812.

Forman BM, Tzameli I, Choi HS, et al. Androstane metabolites bind to and deactivate the nuclear receptor CAR-beta. *Nature*1998; 395(6702):612–5.

Grundy S.M., Cleeman J.I., Daniels S.R., Donato KA, Eckel RH, Franklin BA, Gordon DJ, Krauss RM, Savage PJ, Smith SC Jr, Spertus JA, Costa F. Diagnosis and management of the metabolic syndrome. An American Heart Association/ National Heart, Lung, and Blood Institute scientific statement. *Circulation* -2005.- Vol.112.- P.2735-2752.

Hamblin M, Chang L, Fan Y, Zhang J, Chen YE. PPARs and the cardiovascular system.*Antioxid. Redox. Signal.* 2009;11(6):1415-52.

Herzog, B., Hallberg, M., Seth, A., Woods, A., White, R., and Parker, M.G.,. The nuclear receptor cofactor, receptor-interacting protein 140, is required for the regulation of hepatic lipid and glucose metabolism by liver X receptor. *Mol. Endocrinol.*, 2007, vol. 21, pp. 2687–2697.

Holvoet P, Lee DH, Steffes M, Gross M, Jacobs DR Jr. Association between circulating oxidized low-density lipoprotein and incidence of the metabolic syndrome.*JAMA*2008 May 21;299(19):2287-93.

Issemann I., Green S. Activation of a member of the steroid hormone receptor superfamily by peroxisome proliferators. *Nature*1990;347:645–650.

Kainu, T., Kononen, J., Enmark, E., Gustafsson, J.A., *and*Pelto-Huikko, M.1996. Localization and ontogeny of the orphan receptor OR-1 in the rat brain. *J. Mol. Neurosci.*7*:* 29–39.

Kliewer SA, Goodwin B, Willson TM The nuclear pregnane X receptor: a key regulator of xenobiotic metabolism. *Endocr. Rev.* 2002;23:687–702.

Lalwani ND. Alvares K. Reddy MK. Reddy MN. Parikh I. Reddy JK. Peroxisome proliferator-binding protein: identification and partial characterization of nafenopin-, clofibric acid-, and ciprofibrate-binding proteins from rat liver. *Proc. Natl. Acad. Sci. USA*1987;84:5242–5246.

Landes N, Pfluger P, Kluth D, et al. Vitamin E activates gene expression via the pregnane X receptor. *Biochem. Pharmacol.* 2003;65:269–273.

Lefebvre, P., Chinetti, G., Fruchart, J.-C., and Staels, B., J. Sorting out the roles of PPAR alpha in energy metabolism and vascular homeostasis. *Clin. Invest.*, 2006, vol. 116, pp. 571–580.

Lehmann, J.M., Kliewer, S.A., Moore, L.B., Smith-Oliver, T.A., Oliver, B.B., Su, J.L., Sundseth, S.S., Winegar, D.A., Blanchard, D.E., Spencer, T.A., *et al.*1996. Activation of the nuclear receptor LXR by oxysterols defines a new hormone response pathway. *J. Biol. Chem.*272*:* 3137–3140.

Li H, Wang H. Activation of xenobioticreceptors: driving into the nucleus.*Expert Opin. Drug Metab. Toxicol.* 2010 Apr;6(4):409-26.

Maglich JM, Stoltz CM, Goodwin B, Hawkins-Brown D, Moore JT, Kliewer SA. Nuclear pregnane X receptor and constitutive androstane receptor regulate overlapping but distinct sets of genes involved in xenobiotic detoxification. *Mol. Pharmacol.* 2002; 62:638–646.

Marino, M., di Masi, A., Trezza, V., Pallottini, V., Polticelli, F., Ascenzi, P. Xenosensors CAR and PXR at Work: Impact on Statin Metabolism. *Current Drug Metabolism*, Volume 12, Number 3, 2011 , pp. 300-311(12)

Mathieu, P., Pibarot, P., Despres, J.-P. Metabolic syndrome: the danger signal in atherosclerosis. (2006) Vasc. *Health Risk Manag.*, 2, 285-302.

McLaughlin T., Reaven G., Abbasi F., Lamendola C, Saad M, Waters D, Simon J, Krauss RM. Is there a simple way to identify insulin-resistant individuals at increased risk of cardiovascular disease? *Am. J. Cardiol.* – 2005. – Vol. 96. – P. 399-404.

Michalik, L., Auwerx, J., Berger, J.P., Chatterjee, V.K., Glass, C.K., Gonzalez, F.J., Grimaldi, P.A., Kadowaki, T. , Lazar, M.A., O'Rahilly, S., Palmer, C.N., Plutzky, J., Reddy, J.K., Spiegelman, B.M., Staels, B., and Wahli, W., International Union of Pharmacology. LXI. Peroxisome proliferator-activated receptors. *Pharmacol. Rev.*, 2006, vol. 58, pp. 726–741.

Mnif W, Pascussi JM, Pillon A, Escande A, Bartegi A, Nicolas JC, Cavaillès V, Duchesne MJ, Balaguer P. Estrogens and antiestrogens activate hPXR. *Toxicol. Lett.* 2007;170:19–29.

Moreau, A., Vilarem, M.J., Maurel, P. , and Pascussi, J.M., Xenoreceptors CAR and PXR activation and consequences on lipid metabolism, glucose homeostasis, and inflammatory response. *Mol. Pharm.*, 2008, vol. 5, pp. 35–41.

NCEP-ATP III, *J.A.M.A.*, 2001, vol. 285, pp. 2486– 2497.

Patsouris, D., Mandard, S., Voshol, P.J., Escher, P. , Tan, N.S., Havekes, L.M., Koenig, W. , März, W., Tafuri, S., Wahli, W. , Müller, M., and Kersten, S., PPARalpha governs glycerol metabolism. *J. Clin. Invest.*,2004, vol. 114, pp. 94–103.

Peet, D.J., Janowski, B.A., *and*Mangelsdorf, D.J.1998. The LXRs: A new class of oxysterol receptors. *Curr. Opin. Gen. Dev.*8: 571–575.

Reaven, G.M., Banting lecture 1988. Role of insulin resistance in human disease.*Diabetes*, 1988, vol. 37, pp. 1595–1607.

Reaven G. The metabolic syndrome or the insulin resistant syndromeand Different names, different concepts, and different goals. Endocrinol. *Metable Clin. North. Am.* – 2004. – Vol. 33. – P. 283-303.

Reddy J. Svoboda D. Azarnoff D. Microbody proliferation in liver induced by nafenopin, a new hypolipidemic drug: comparison with CPIB. *Biochem. Biophys. Res. Commun.*1973;52:537–543.

Reddy JK. Hepatic microbody proliferation and catalase synthesis induced by methyl clofenapate, a hypolipidemic analog of CPIB. *Am. J. Pathol.* 1974;75:103–118.

Reddy JK. Krishnakantha TP. Hepatic peroxisome proliferation: induction by two novel compounds structurally unrelated to clofibrate. *Science* (New York) 1975;190:787–789.

Reddy, J.K., and Rao, M.S., Lipid metabolism and liver inflammation. II. Fatty liver disease and fatty acid oxidation. *Am. J. Physiol. Gastrointest. Liver Physiol.*, 2006, vol. 290, pp. G852–G858.

Robinson E, Grieve DJ. Significance of peroxisome proliferator-activated receptors in the cardiovascular system in health and disease. *Pharmacol. Ther.* 2009 Jun;122(3): 246-63

Sassolas, A., Vincent, M., Benzoni, D., and Sassard, J., Plasma lipids in genetically hypertensive rats of the Lyon strain. *J. Cardiovasc. Pharmacol.*, 1981, vol. 5, pp. 1008–1014.

SchultzJR, TuH, LukA, RepaJJ, MedinaJC, LiL, SchwendnerS, WangS, ThoolenM, MangelsdorfDJet al.2000Role of LXRs in control of lipogenesis. *Genes and Development*142831–2838.

Semple RK. Meirhaeghe A. Vidal-Puig AJ. Schwabe JW. Wiggins D. Gibbons GF. Gurnell M. Chatterjee VK. O'Rahilly S. A dominant negative human peroxisome proliferator-activated receptor (PPAR){alpha} is a constitutive transcriptional corepressor and inhibits signaling through all PPAR isoforms. *Endocrinology*. 2005; 146:1871–1882.

Shimamoto K, Ura N. Mechanisms of insulin resistance in hypertensive rats.*Clin. Exp. Hypertens*. 2006;28(6):543-52.

Sueyoshi T, Kawamoto T, Zelko I, Honkakoski P, Negishi M 1999The repressed nuclear receptor CAR responds to phenobarbital in activating the human CYP2B6 gene. *J. Biol. Chem.*274:6043–6046.

Swales K, Negishi M. CAR, driving into the future.*Mol. Endocrinol.* 2004;18(7):1589-98.

Tabb MM, Sun A, Zhou C, et al. Vitamin K2 regulation of bone homeostasis is mediated by the steroid and xenobiotic receptor SXR. *J. Biol. Chem.* 2003;278:43919–43927.

Taketa K, Matsumura T, Yano M, Ishii N, Senokuchi T, Motoshima H, Murata Y, Kim-Mitsuyama S, Kawada T, Itabe H, Takeya M, Nishikawa T, Tsuruzoe K, Araki E. Oxidized low density lipoprotein activates peroxisome proliferator-activated receptor-alpha (PPARalpha) and PPARgamma through MAPK-dependent COX-2 expression in macrophages.*J. Biol. Chem.*2008 Apr 11;283(15):9852-62.

Tien ES, Negishi M. Nuclear receptors CAR and PXR in the regulation of hepatic metabolism. *Xenobiotica*. 2006 Oct-Nov;36(10-11):1152-63.

Timsit YE, Negishi M. CAR and PXR: the xenobiotic-sensing receptors. *Steroids*. 2007;72:231–246.

Tobin, K.A., Steineger, H.H., Alberti,S., Spydevold, O., Auwerx, J., Gustafsson, J.A., and Nebb, H.I., Cross-talk between fatty acid and cholesterol metabolism mediated by liver X receptor-alpha. *Mol. Endocrinol.*, 2000, vol. 14, pp. 741–752.

Ueda A, Hamadeh HK, Webb HK, Yamamoto Y, Sueyoshi T, Afshari CA, Lehmann JM, Negishi M. Diverse roles of the nuclear orphan receptor CAR in regulating hepatic genes in response to phenobarbital. *Mol. Pharmacol.*, 2002,61*:1–6.*

Van der Leij, F.R., Bloks, V.W., Grefhorst, A., Hoekstra, J., Gerding, A., Kooi, K., Gerbens, F. , te Meerman, G., and Kuipers, F. , Gene expression profiling in livers of mice after acute inhibition of beta-oxidation. *Genomics*, 2007, vol. 90, pp. 680–689.

Wada Taira , Jie Gao and Wen Xie PXR and CAR in energy metabolism. *Trends in Endocrinology and Metabolism* 2009,Volume 20, Issue 6, Pages 273-279

Willson TM, Kliewer SA. PXR, CAR and drug metabolism. *Nat. Rev. Drug Discov*. 2002;1:259–266.

Willy, P.J., Umesono, K., Ong, E.S., Evans, R.M., Heyman, R.A., *and* Mangelsdorf, D.J.1995. LXR, a nuclear receptor that defines a distinct retinoid response pathway. *Genes and Dev.*9*:* 1033–1045.

Woods, C.G., Heuvel, J.P., and Rusyn, I., Genomic profiling in nuclear receptor-mediated toxicity. *Toxicol. Pathol.*, 2007, vol. 35, pp. 474–494.

Xiao L, Xie X, Zhai Y. Functional crosstalk of CAR-LXR and ROR-LXR in drug metabolism and lipid metabolism. *Adv.DrugDeliv.Rev.* 2010 Oct 30;62(13):1316-21.

Yu, S-M., Kang, Ya-F., Chen, C.-C., and Teng, C.-M., Effects of dicentrine on haemodynamic, plasma lipid, lipoprotein level and vascular reactivity in hyperlipidaemic rats. *Br. J. Pharmacol.*, 1993, vol. 108, pp. 1055–1061.

Zhou J, Zhai Y, Mu Y,Gong H, Uppal H, Toma D, Ren S, Evans RM, Xie W. A novel pregnane X receptor-mediated and sterol regulatory element-binding protein-independent lipogenic pathway. *J. Biol. Chem.* 2006, vol. 281, pp. 15013–15020.

Zhou J, Febbraio M, Wada T, Zhai Y, Kuruba R, He J, Lee JH, Khadem S, Ren S, Li S, Silverstein RL, Xie W. Hepatic fatty acid transporter Cd36 is a common target of LXR, PXR, and PPARγ in promoting steatosis. *Gastroenterology*. 2008;134:556–567.

In: Dyslipidemia ISBN: 978-1-62100-917-7
Editors: Z. Ačimovič et.al, pp. 199-215

Chapter 7

ENDOTHELIAL LIPASE: ITS ROLE IN THE METABOLISM OF HDL- CHOLESTEROL, THE RELATIONSHIP TO CARDIOVASCULAR RISK AND THERAPEUTIC OPPORTUNITIES

D. Agapakis, S. Fotiadis, C. Savopoulos, and A. I. Hatzitolios

First Propedeutic Internal Medicine Department, AHEPA Hospital, Medical School, Aristotle University of Thessaloniki, Thessaloniki, Greece

ABSTRACT

The endothelial lipase (EL) is an enzyme with lipolytic activity against phospholipids of high density lipoprotein-cholesterol (HDL) in plasma.

The EL is synthesized by vascular endothelial cells. In particular, it is synthesized mainly in heart, muscle, and adipose tissue and then transported to the luminal surface of endothelial cells where it hydrolyzes lipoprotein triglycerides and phospholipids. In addition, the endothelial-derived EL can be a direct regulator of HDL metabolism increasing the catabolism of HDL particles, thereby reducing the levels and the size of the tank and impacting recycling. B

Both experimental in vitro and in vivo studies have revealed remarkable information with regard the physiological role of EL in inflammatory conditions and its potential role in modulating lipoprotein metabolism under inflammatory conditions, including atherosclerosis. EL also seems to have local actions in the vascular wall and the regulation of circulating lipoproteins, and it seems to be associated with established and novel cardiovascular risk factors like hypertension, diabetes, dyslipidemia, obesity, and leptin, which potentially contributes to the progress of atherosclerosis and therefore could be regarded as a biochemical marker of atherosclerosis.

The EL inhibition is considered to be an attractive intervention target in order to favorably modulate the atherosclerosis process by improving the HDL levels.

This article reviews the effect of EL on the metabolism of HDL-C, its association with cardiovascular risk factors and the potential perspective to arrest or even improves atherosclerosis by EL inhibition.

INTRODUCTION

From large multi-center clinical intervention trials of the last decade it has been demonstrated the inverse relationship between levels of cholesterol in high density lipoprotein (HDL) and cardiovascular risk confirming what has been observed in previous epidemiological and experimental studies, and thus further supporting the antiatherogenic role of the HDL [1-4]. It has also been found that low levels of HDL are the most important independent risk factor for morbidity and mortality in advanced heart failure, compared to total cholesterol (T-C) and low density lipoprotein (LDL) [5]. Moreover, greater predictive value in preventing cardiovascular risk is offered when HDL is included instead of using T-C or LDL by themselves. [6].

Therefore HDL increase continues to be an attractive therapeutic challenge aiming to atherosclerosis prevention and novel treatments are attempted to be developed. With regard HDL action, therapeutic intervention focuses at different steps including the transport of cholesterol from peripheral tissues to the liver, inhibition of oxidation of LDL, arresting vascular inflammation and improving endothelial antithrombotic properties. Both the function and structure of HDL-particles contribute to atheroprotecive actions. Among the factors that shape the metabolism of HDL, endothelial lipase (EL) is recognized as an important one.

Endothelial Lipase: Its Role in the Metabolism of HDL

This is a new member of of the lipase enzyme family, a glycoprotein 68 kDa encoded by a gene on chromosome LIPG 18q21.1, which was discovered in the relatively recently [8]. The EL is characterized by lipolytic activity of phospholipase A1 which has almost exclusively as a substrate phospholipids of HDL-particles which are hydrolyzed in position sn-1, in contrast to the lipoprotein, liver and pancreatic lipase which are predominantly triglyceride lipases [9]. It is produced in a unique way mainly by endothelial cells and to a lesser extent by macrophages and smooth muscle cells. The tissue expression is mainly in liver, thyroid, lungs, kidneys, testes and placenta. After production, the EL binds to proteoglycans on the endothelial surface initiating its local action [9]. Two potential lipid binding sites and four heparin or proteoglycans have been identified. EL promotes, the hydrolysis of phospholipids leading to the creation of small, dense HDL-particles which are structurally unstable, thus facilitating the release of lipid poor apolipoprotein (Apo) A-I.

Overexpression of EL in mice was associated with reduced levels of HDL and Apo AI due to increased catabolism and excretion by the kidneys [10-11]. The hydrolysis of phospholipids result non-esterified fatty acids that are received and incorporated into cellular phospholipids and triglycerides increasing the total amount of cellular lipids, while is suppresses the rate of synthesis of fatty acids. The remaining residual cholesterol in HDL particles is particularly sensitive to selective hepatic uptake, which is greatly increased, through receptor SR-B1 [10, 12]. On the contrary, in experimental models where EL activity was abolished, an increase in the quantity and size of HDL-particles was observed even under nutrition conditions with foods rich in cholesterol, which are highly atherogenic [13 -14].

Thus, the fluctuating lipolytic activity of EL remodels and modifies the phospholipid content of the HDL particles, through dose-related increase or decrease of catabolic rate [10, 13]. Besides the above enzymatic activities, the EL can also enhance cell attachment and processing of HDL-particles to the endothelium through heparinic sulfate proteoglycan assisted pathaways. [15]. Enzymatic and non-enzymatic functions of EL appear to be interrelated synergistic.

In conclusion, in vitro and in vivo studies have documented that EL facilitates through two ways the catabolism of HDL, modifying its concentration and composition, as an important negative regulator of HDL concentration [13, 16]. It seems therefore that EL might prove a new promising therapeutic target.

THE ROLE OF EL IN REVERSE CHOLESTEROL TRANSPORT (RCT)

The RCT is the key atheroprotective mechanism of HDL, including the efflux of cholesterol from intracellular reservoirs in peripheral cells and foam cells of atherosclerotic plaques and its transfer to the liver [17].

This function includes mainly the one-way efflux of free cholesterol from intracellular cholesterol pool from the atherosclerotic lesions of the macrophage foam cells, toward lipid poor Apo AI by ATP binding-cassette transporter 1 (ABCA1) where they are changed to nascent discoidal HDL.

A two-way flow of free cholesterol between cells and HDL via receptor SR-B1, whose ligands are rich in phospholipids lipoproteïnes [18-19]. Through the actions of lecithin cholesterol acyltransferase (LCAT) first and then of the protein transport of phospholipids (PLTP) are generated the mature spherical HDL-3 initially and successively the large HDL-2 particles which have antiatherogenic properties. Both forms of HDL are substrates for the hepatic and endothelial lipase. The large HDL-2 either interact with VLDL and LDL cholesterol ester exchanging triglycerides via the enzyme cholesterol ester transfer (CETP), or attribute esterified cholesterol in liver cells through hepatic receptor SR-B1 to be further metabolized and excreted as bile acids, while the same HDL-2 will be converted back to small HDL-3, which also have antiatherogenic properties [20]. Noteworthy that the ability of HDL particles to employ and pay lipids is inversely proportional to the amount of fat they contain.

The specific effect of EL on intracellular cholesterol efflux mediated by HDL is not supported by many studies and not always leads to clear conclusions. It is considered however reasonable to assume that the inhibition of EL strengthens the mechanism of RCT in vivo [21]. It has been suggested that depletion of phospholipids in HDL induced by the EL leads to decreased affinity of the first with the cell membrane and reduced ability to efflux cholesterol to HDL. Indeed, studies in vitro, shows that overexpression of EL was associated with changes in chemical composition and physical properties of HDL, which had resulted in reduced binding capacity of HDL to SR-B1 receptor in peripheral cells and thereby decreased the selective free cellular cholesterol uptake through this receptor [22]. Moreover, the subsequent reduction in the nucleus of HDL-particles due to loss of phospholipids, results in their structural instability, as has been mentioned, and the released Apo AI will interact with the ABCA1-mediated cholesterol efflux. This was found in the experimental study of Qiu and Hill where overexpression of EL in macrophages altered the composition of phospholipids of cell membranes and produced Lyso-phosphatidylcholine (lyso-PC) that has the ability to stimulate the binding of Apo AI with the ABCA1 receptor resulting in increased efflux of cholesterol, through both enzyme and non-enzymatic mechanisms of EL. While the suppression of EL has the opposite effect [23]. These results suggest that EL can promote the inhibition of cholesterol efflux from macrophages. Similarly, Yancey et al. observed that overexpression of EL caused a 63% increase of cholesterol efflux through the ABCA1 receptor and reduced cholesterol efflux through the SR-B1 receptor by 90%, probably due to increased production of pre-beta HDL in serum [24]. Instead, in his study Gauster et al. didn't find any increase of cholesterol efflux through ABCA1 under similar circumstances. In a recent and very important study of Braun JR et al. it was observed large increase in cholesterol efflux capacity of HDL in cases of shortage of ($EL^{-/-}$) mice [25]. These findings highlight the need for more studies on the in vivo effects of EL on RCT and that overexpression or lack of EL seem to change the degree of involvement of free cholesterol efflux pathways, via activation of different cellular receptors, presumably in an effort to maintain the homeostasis of cholesterol.

On the other hand, the EL acts in the second phase of RCT hepatic uptake of cholesterol ester in HepG2 cells via SR-B1 and tethering HDL-particles [12, 26]. In particular, it was found that the hepatic uptake of cholesterol requires the formation of HDL-particles by hepatic expression of EL [12]. Nevertheless, the total body cholesterol excretion, which is the last stage of RCT following hepatic metabolism and the formation of bile acids, remains unchanged in mice with $EL^{-/-}$ despite the increasing concentration of HDL in plasma [21, 25].

In conclusion, it can be said that the inhibition of EL results in increased levels of HDL without interfering with the hepatic uptake of cholesterol, and this is a positive contribution to the process RCT with the subsequent reduction in atheroma in vascular lesions. However, additional clinical studies are needed in order to quantify the magnitude of cholesterol efflux capacity in conditions of changing levels of EL.

EL AS A RISK FACTOR FOR ATHEROSCLEROSIS

Several studies have supported the positive role of EL in the formation and evolution of the atherosclerotic lesion. Thus, it is found that the EL can be expressed in endothelial cells, smooth muscle cells of the intima and also in neo-vascular network of atherosclerotic lesions in coronary arteries, suggesting that it plays an active role even in early atherosclerotic lesions [11]. Presumably by the action of EL released free fatty acids and phospholipids are offered as a source of energy for the functioning of endothelial and smooth muscle cells of blood vessels [27]. Recently, Huang et al. found higher concentration of EL in plasma of people with atherosclerosis compared with healthy individuals [28]. The proatherogenic properties of EL are also established from studies where the overexpression of EL was found to enhance the adhesion of monocytes to the vascular wall, thereby promoting atherosclerosis [29-30].On the other hand, the work of Ishida et al. examined atherosclerotic models in transgenic mice with functional deletion of EL (apo $E^{-/-}$, $LIPG^{-/-}$), and found a reduction in the size of atherosclerotic lesions, reduction of vascular infiltration of macrophages and increased levels of HDL [31]. Moreover, there are further data which support the association of EL with inflammatory processes. In the experimental study of Yasuda et al. it was demonstrated that the cytokines TNF-a and IL-1b induce the expression of mRNA of EL endothelial cells as enhancing the mechanical forces involved in vascular diseases [32-33]. Similar studies in vitro in macrophage cell cultures and in vivo also found positive relationship between pro-inflammatory gene and the expression of EL in vascular endothelium, which in turn affects the synthesis of cytokines sustaining a vicious cycle [30, 34]. Different results emerged from the work of Ko et al. in hyperlipidemic mouse $apoE^{-/-}$ and $LDLR^{-/-}$ where no change was observed in atherosclerosis, despite an increase in HDL as result of EL's inactivation [35]. However, the experimental model chosen is extremely atherogenic and it is uncertain what kind of extrapolations can be made from this study.

On the other hand, the study of Ahmed et al. found that the intact activity of EL on HDL is essential for activation of the nuclear receptor PPAR, and this activation represents the main mechanism for the expression of anti-inflammatory action of HDL ie inhibition of vascular adhesion of leukocytes [36]. However, activation of PPAR may only be an indicator of the intensity of the inflammatory process and the role of EL probably emphasizes the link between abnormal action and reactive response. Some anti-atherogenic properties of EL suggest the work of Brown et al where the lack of EL and hepatic lipase resulted in the formation of small dense LDL [25]. In the same vein, the Broedl et al. showed that in lipidemic mice the overexpression of EL caused an increase in clearance of Apo B lipoproteins [31, 37]. While the lack of EL was associated with increased levels of VLDL / LDL as a result of mild lipolytic action of the TRG and increasing residence time of VLDL in plasma [35]. However, the impact on human's EL on Apo B lipoproteins is limited and the reciprocal increase in anti-atherogenic and atherogenic lipoproteins in LIPG^{-} /$^{-}$, Apo E^{-}/$^{-}$ mice do not appear to be balanced, and it is associated with dramatic reduction of atherosclerotic volume by 70% tilting the balance towards the atheroprotective increased levels of HDL [31].

With regard other mechanisms of HDL atheroprotection, it has been shown that the EL promotes engagement and recruitment of non-oxidized and oxidized LDL in macrophages [23]. It has been shown, that the inhibition of EL enhances all atheroprotective mechanisms of HDL which have been shown both in vitro and vivo (efflux of cholesterol, anti-inflammatory, antioxidant, PON-1, PAF-AH, antithrombotic) [38].

Finally, infection increases the expression of EL and in this way modifies the lipopolysaccharide-neutralizing capacity of HDL (removing phospholipids). Also, mice EL -/- showed increased ability of HDL to inhibit the secretion of TNF-α which stimulates the lipopolysaccharide and injected HDL from these animals in wild mice infected with endotoxin showed a longer survival [38].

This behavior of HDL, as a key component of innate immunity should be considered since torcetrapib (inhibitor of CETP) led to a true increase of HDL but increased also the frequency of deaths secondary to infections [39].

In conclusion, the EL is involved in all pathophysiological mechanisms of atherosclerosis associated with low levels of HDL, while its targeted inhibition leads to atheroprotectve HDL-particles and interesting immunological activities.

EL Correlation with Cardiovascular Disease and Risk Factors

These features of the EL have been confirmed in clinical studies and have been linked to levels of EL with established and novel cardiovascular risk factors, the metabolic syndrome and the progress of atherosclerosis. Particularly in the study of Badellino et al. asymptomatic individuals with a family history of premature coronary heart disease, an increase in EL mass was positively correlated with all risk factors, i.e. obesity (waist circumference, body mass index>30), hypertension, fasting glucose, smoking, lack of exercise, insulin resistance, the hypertriglyceridemia, the concentration of Apo B, the LDL in women, and calcification of the coronary artery which is a well known indicator of subclinical atherosclerosis [40]. A similar correlation has also been found between increased EL and increased visceral obesity and proatherogenic lipid profiles associated with it (T-C, VLDL, TRG, Apo B) [41]. Interestically, it appears that the EL is an alternative way of recruiting FFA from adipose tissue in mice $LPL^{-/-}$, with serious effects on the metabolism of lipoproteins [42]. Additionally, it has been shown that inflammatory markers such as hs-CRP, sTNFr2, IL-6, sICAM-1 and leptin correlated positively and directly with EL. Thus, overexpression of EL in obesity and metabolic syndrome may be explained by the proinflammatory state where inflammatory cytokines released by fat cells [34]. In contrast, negative correlation was found between EL and adiponectin, which is found at low levels in obesity and it is known to inhibit the TNF-a induced EL. This effect is achieved through inhibition of the nuclear factor-KB (NF -kB) a key initiator of secretion of EL [34]. The study of Yoko et al. identified a clear negative correlation between levels of EL with the levels of HDL, in patients who had developed cardiovascular disease [43]. On diabetic patients, the recent study by Sammy et al. showed that the EL is expressed in excess (Diabetes Mellitus can be regarded as a state of subclinical inflammation), although the impairment of cholesterol efflux mechanism was associated with very low HDL levels in these patients [44]. Also, a small study found EL in vitro to be proportional to the extracellular glucose concentration, but not in vivo [45]. While a larger study, found increased concentration of EL in diabetics who were under oral anti-diabetic treatment but not in insulin treated patients [46]. It appears that insulin alters the expression of EL. Also, a positive association of EL with CRP in diabetic and non-diabetic patients has been observed, but not in insulin treated patients [46]. Finally, a clinical study of Fujii et al. assessed the role of EL in dialysis patients and found correlation between EL, hypoalbuminemia and inflammation in these patients [47].

In conclusion, there is growing evidence that the EL is closely linked to the factors and markers of cardiovascular risk and plays an active role in shaping the profile of atherogenic dyslipidemia that characterizes the metabolic syndrome.

GENE POLYMORPHISMS OF EL AND HDL

The gene polymorphisms of LIPG are associated with functional changes in the action of EL and participate in shaping levels of HDL [48].They can also play an important role in the extent of morbidity associated with elevated levels of EL and the therapeutic response to pharmacological interventions. For these reasons, these polymorphisms have been the subject of intensive research and the most common mononoukleotide polymorphism 584 C/T produces a variety Thr111Ile, is the focus of interest of several researchers. Ma et al. found a significant correlation between the correspondence of 584 C/T and HDL levels but no difference in the development of CHD [49]. Tang et al. in a Chinese population found that the T allele correspondence of the same mononoukleotide polymorphism 584 C/T offers cardiovascular protection [50]. Unlike studies in Caucasian healthy populations showed no association of 584 C/T with HDL levels or CHD [51]. While in Japanese patients with AMI a positive correlation with the polymorphism was observed, regardless of HDL [52]. Also, the polymorphism 584 C/T was associated with a higher incidence of retinopathy in diabetic patients with lower levels of LDL, mild increase in HDL and arterial blood pressure too [53]. The diversification of the distribution of the gene variant of LIPG according to race and ethnicity and the differing impact of environment on the functional expression of EL in the populations studied (such as association with smoking, type of diet) produces inconsistent results of investigations and make it difficult to generalize the findings. More extensive investigations are necessary to confirm any correlation between the gene of EL with the levels, the type of HDL particles and cardiovascular disease.

FACTS AND CHALLENGES IN THE THERAPEUTIC TARGETING OF EL

These findings of EL and its inverse relationship with HDL shape the challenge to design new methods and therapies aimed at inhibiting EL as a tool to increase HDL and the subsequent reduction cardiovascular risk. .

So far there are no specific drugs to inhibit selectively the expression of EL. There are available easy to use biochemical methods to determine the change in the activity of the enzyme [54-55]. From biotechnology and biochemistry are under development substances with inhibition activity. The sulfonylfuran urea 1 is one such substance potentially suitable to inhibit the EL in a selective manner and is under development [56]. Sulpharaphan (SFN) which is a naturally occurring isothiocyanate present in cruciferous vegetables, has important antioxidant and anti-inflammatory effects, *inhibits EL expression through NF-κB in endothelial cells* [57]. Still, the boronic acid may seem to be the basis for the design of new drugs since in vivo testing has shown to inhibit EL and to increase HDL by about 30% [58].

Similar efforts are also in the field of molecular pharmacology of the composition antisense nucleotides to the intervention and modification of gene expression of EL, while developing methods for future diagnostic and therapeutic approach (kits) in patients who develop morbidity related to the EL [59]. From already used lipid-lowering drugs, statins cause a small increase in levels of HDL. However, their contribution is important mainly because qualitatively alter the dysfunctional HDL-proathrerogenic occurring in conditions of metabolic syndrome, diabetes, CHD, infections. Only two studies exist to assess the effect of statins on levels of EL. In the in vitro study of Qiu and Hill evaluated the effect of administration of atorvastatin and simvastatin in the expression of EL in macrophages in human cell cultures. Significant dose-dependent reduction of EL mRNA was found in macrophages via mechanisms of inhibition of HMG-CoA reductase and the NF-kB [60]. In a clinical study of Kojima Y et al. the administration for six months of pitavastatin in hypercholesterolaemic patients, reduced the levels of EL by 14%, increased HDL levels by 11% and significantly increased the size of HDL-particles by 12%, supporting the view of pleiotropic action of statins. Furthermore, in the same study it was found that Rho A is an important factor pathway in regulating the expression of EL in endothelial cells [43]. The effect of fibrates and nicotinic acid on the expression of EL there are currently no available data. However, it is known that fibrates are PPARa activators. The activation of these receptors causes increased expression of proprotein convertase PC 5/6 and furin, which in turn causes proteolytic cleavage of EL on the cell surface [61-62]. Thus, it seems logical that the beneficial effects of these drugs on HDL, through a specific regulatory pathway also include reducing the EL. The niacin is the most effective lipid lowering agent for increasing HDL, has properties of PPAR-gamma agonist and exhibits pleiotropic actions such as increased adiponectin and reduction of CRP. Finally, pioglitazone also acts as a PPAR-gamma activator and has the specificity to trigger factor IkBα in vitro and in vivo in experimental models. This factor is an inhibitor of NF-kB, which is positively correlated with the induction of expression of EL and the VCAM-1. Consequently, pioglitazone is expected to reduce the levels of EL [63]. Overall, it appears that the PPARs are involved in the pathophysiology of HDL homeostasis through the regulation of secretion of EL.

Conclusion

In conclusion, the EL acts as a regulator of HDL levels and is associated with an inverse relationship. It has proatherogenic properties, in particular vascular expression and may interfere with the atherosclerotic process through direct local action in the vascular wall and through the control of circulating lipoproteins. The EL is positively correlated with cardiovascular risk factors and could be considered a biochemical marker of inflammatory and atherosclerotic process. In clinical practice, it appears that certain statins, including simvastatin, atorvastatin and pitavastatin may partially inhibit this action of El and at present we look forward to novel treatments which will target EL inhibition.

References

[1] Paramsothy, P; Knopp, RH; Bertoni, AG; Blumenthal, RS; Wasserman, BA; Tsai, MY; Nathan, TR; Wong, D; Heckbert, SR. Association of Combinations of Lipid Parameters With Carotid Intima-Media Thickness and Coronary Artery Calcium in the MESA (Multi-Ethnic Study of Atherosclerosis) *J. Am. Coll. Cardiol.* 2010; 56;1034-1041.

[2] Taylor, AJ; Villines, TC; Stanek, EJ; Pharm, D; Devine, PJ; Len Griffen, Miller, M; Weissman, NJ; Turco, M. Extended-Release Niacin or Ezetimibe and Carotid Intima–Media Thickness. *N Engl J Med* 2009; 361:2113-2122.

[3] Taylor, AJ; Sullenberger LE; Lee HJ; Lee JK; Grace KA. Arterial Biology for the Investigation of the Treatment Effects of Reducing Cholesterol (ARBITER) 2: a double-blind, placebo-controlled study of extended-release niacin on atherosclerosis progression in secondary prevention patients treated with statins. *Circulation* 2004, Dec 7; 110(23):3615.

[4] Sanossian, N; Saver, JL; Navab, M; Ovbiagele, B. High-Density Lipoprotein Cholesterol: An Emerging Target for Stroke Treatment. *Stroke* 2007; 38;1104-1109.
[5] Mehra, MR; Uber, PA; Lavie, CJ; Milani, RV; Park, MH; Ventura HO. High-density lipoprotein cholesterol levels and prognosis in advanced heart failure. *J Heart Lung Transplant.* 2009, Sep; 28(9):876-80.
[6] Millán, J; Pintó, X; Muñoz, A; Zuñiga, M; Rubiés-Prat, J; Pallardo, LF; Massana, L; Mangas, A; Hernández-Mijares, A; González-Santos, P; Ascaso, JF; Predo-Botet, J. Lipoprotein ratios: Physiological significance and clinical usefulness in cardiovascular prevention. *Vasc Health risk Manag.* 2009; 5:757-65.
[7] Assmann, G; Gotto, AM. HDL Cholesterol and Protective Factors in Atherosclerosis. Circulation 2004; 109:III-8 – III-14.
[8] Jaye, M; Lynch, KJ; Krawiec, J; Marchadier, D;, Maugeais, C; Doan, K; South, V; Amin, D; Perrone, M; Rader, DJ. A novel endothelial-derived lipase that modulates HDL metabolism. *Nat. Genet.* 1999; 21:424-428.
[9] McCoy, MG; Sun, GS; Marchadier, D; Maugeais, C; Glick, JM; Rader, DJ. Characterization of the lipolytic activity of endothelial lipase. *J Lipid Res* 2002; 43:921-9.
[10] Maugeais, C; Tiefge, UJ; Broedl, UC; Marchadier, D; Cain, W; McCoy, MG; Lund-Katz, S; Glick, JM; Rader, DJ. Dose-dependent acceleration of high-density lipoprotein catabolism by endothelial lipase. *Circulation* 2003; 108:2121-2126.
[11] Yasuda, T; Ishida, T; Rader, DJ. Update on the role of endothelial lipase in high-density lipoprotein metabolism, reverse cholesterol transport, and atherosclerosis. *Circ J.* 2010; 74:2263–70.
[12] Nijstad, N; Wiersma, H; Gautier, T; van der Giet, M; Maugeais, C; Tietge, UJ. Scavenger receptor BI-mediated selective uptake is required for the remodeling of high density lipoprotein by endothelial lipase. *J Biol Chem* 2009; 284: 6093 – 6100.
[13] Jin, W; Millar, JS; Broedl, U; Glick, JM; Rader, DG. Inhibition of endothelial lipase causes increased HDL cholesterol levels in vivo. *J Clin Invest* 2003 Feb; 111(3):357-62.
[14] Williams, PT; Feldman, DE. Prospective study of coronary heart disease vs. HDL2, HDL3, and other lipoproteins in Gofman's Livermore Cohort. *Atherosclerosis* 2011; 214(1):196-202.
[15] Fucki IV; Blanchard, N; Jin, W; Marchadier, DH; Millar, JS; Glick, JM; Rader, DJ. Endogenously produced endothelial lipase enhances binding

and cellular processing of plasma lipoproteins via heparin sulfate proteoglycan-mediated pathway. *J Biol Chem* 2003 Sep 5; 278(36):34331-8.

[16] Cohen, JC. Endothelial lipase: direct evidence for a role in HDL metabolism. *J Clin Invest.* 2003 February 1; 111(3):318-321.

[17] Tall, AR. Cholesterol efflux pathways and other potential mechanisms involved in the athero-protective effect of high density lipoproteins. *J. Intern. Med.* 2008 Mar; 263(3):256-73.

[18] Von Escardstein A, Nofer JR, Assmann G. HDL and arteriosclerosis: role of cholesterol efflux and reverse cholesterol transport. *Arterioscler Thromb vasc Biol* 2001; 21:13-27

[19] de la Llera-Moya, M; Rothblat, GH; Connelly, MA; Kellner-Weibel, G; Sakr, SW; Phillips, MS; Williams, DL..Scavenger receptor BI (SR-BI) mediates free cholesterol flux independently of HDL tethering to the cell surface. *J. Lipid Res.* 1999; 40(3):575-580.

[20] Kontush, A; Chapman, GM. Antiatherogenic small, dense HDL-guardian angel of the arterial wall? *Nat clin Pract Cardiovasc Med* 2006; 3(3):144-153.

[21] Annema, W; Tietge, U. Role of Hepatic Lipase and Endothelial Lipase in High-Density Lipoprotein–Mediated Reverse Cholesterol Transport *Curr Atheroscler Rep* 2011; 13(3):257–265.

[22] Gauster, M; Oskolkova, OV; Innerlohinger, J; Glatter, O; Knipping, G; Frank, S. Endothelial lipase-modified high-density lipoprotein exhibits diminished ability to mediate. *Biochem. J.* 2004; 382: 75–82.

[23] Qiu, G; Hill, JS. Endothelial Lipase Promotes Apolipoprotein AI–Mediated Cholesterol Efflux in THP-1 Macrophages. *Arterioscler Thromb Vasc Biol.* 2009; 29:84-91.

[24] Yancey, PG; Kawashiri, MA; Moore, R; Glick, JM; Williams, DL;Connelly, MA; Rader, DJ; Rothblatet, GH. In vivo modulation of HDL phospholipid has opposing effects on SR-BI- and ABCA1-mediated cholesterol efflux. *J Lipid Res.* 2004; 45:337-46.

[25] Brown, RJ; Lagor, WR; Sankaranaravanan, S; Yasuda, T; Quertermous, T; Rothblat, GH; Rader, DJ. Impact of combined deficiency of hepatic lipase and endothelial lipase on the metabolism of both high-density lipoproteins and apolipoprotein B-containing lipoproteins. *Circ Res.* 2010; 107:357-64.

[26] Strauss, JG; Zimmermann, R; Hrzenjak, A; Zhou, Y; Kratky, D; Levak-Frank, S; Kostner, GM; Zechner, R; Frank, S. Endothelial cell-derived lipase mediates uptake and binding of high-density lipoprotein (HDL)

particles and the selective uptake of HDL-associated cholesterol esters independent of its enzymic activity. *Biochem J* 2002 Nov 15; 368(Pt 1):69-79.

[27] Azumi, H; Hirata, K; Ishida, T; Kojima, Y; Rikitake, Y; Takeuchi, S; Inoue, N; Kawashima, S; Hayashi, Y; Itoh, H; Quertermous, T; Yokoyama, M. *Cardiovascular Research* 2003; 58:647-654.

[28] Huang, J; Qian, H-Y; Li, Z-z; Zhang, Z-M; Wang, S; Tao, Y; Gao, Y-L; Yin, C-Q; Que, B; Sun, T; Zhao, Z-Y; Li, Z. Role of endothelial lipase in atherosclerosis. *Translational research* 2010; 156(1):1-6.

[29] Shimokawa, Y; Hirata, K; Kojima, Y; Iwai, K; Ishida, T; Hirase, T; Inoue, N; Kawashima, S; Yokoyama, M. *Circ J.* 2003; 67(suppl 1):220

[30] Kojma, Y; Hirata, K; Ishida, T; Shimokawa, Y; Inoue, N; Kawashima, S; Quertermous, T; Yokoyamaet, M. Endothelial lipase modulates monocyte adhesion to the vessel wall: A potential role in inflammation. *J Biol Chem*. 2004; 279(52):54032-54038.

[31] Ishida, T; Choi, SY; Kundu, RK; Spin, J; Yamashita, T; Hirata, K; Kojima Y; Yokoyama, M; Cooper, AD; Quertermous, T. Endothelial lipase modulates susceptibility to atherosclerosis in apolipoprotein-E deficient mice. *J Biol Chem*. 2004 Oct 22; 279(43):45085-92.

[32] Yasuda, T; Hirata, K; Ishida, T; Kojima, Y; Tanaka, H; Okada, T; Quertermous, T; Yokoyama, M. Endothelial lipase is increased by inflammation and promotes LDL uptake in macrophages *J. Atheroscler. Thromb*. 2007; 14:192-201.

[33] Choi, SY; Hirata, K; Ishida, T; Quertermous, T; Cooper, AD. Endothelial lipase: a new lipase on the block. *J Lipid Res*. 2002; 43:1763-1769.

[34] Badellino, KO; Wolfe, ML; Reilly, MP; Rader, DJ. Endothelial lipase is increased in vivo by inflammation in humans .*Circulation* 2008; 117(5):678-685

[35] Ko, KW; Paul, A; Ma, K; Li, L; Chan, L. Endothelial lipase modulates HDL but has no effect on atherosclerosis development in apoE−/− and LDLR−/− mice. *J Lipid Res* 2005; 46:2586-2594

[36] Ahmed, W; Orasanu, G; Nehra, V; Asatryan, L; Rader, DJ; Ziouzenkova, O; Plutzky, J. High-Density Lipoprotein Hydrolysis by Endothelial Lipase Activates PPARa : A Candidate Mechanism for High-Density Lipoprotein-Mediated Repression of Leukocyte Adhesion. *Circ Res*. 2006 Mar 3; 98(4):490-8.

[37] Broedl, UC; Maugeais, C; Millar, JS; Jin, W; Moore, RE; Fuki, IV; Marchadier, D; Glick, JM; Rader, DJ. Endothelial lipase promotes the

catabolism of ApoB-containing lipoproteins. *Circ Res* 2004; 94:1554-1561.

[38] Hara, T; Ishida, T; Kojima, Y; Tanaka, H; Yasuda, T; Shinohara, M; Toh, R; Hirata, KI. Targeted deletion of endothelial lipase increases HDL particles with anti-inflammatory properties both in vitro and in vivo. *J. Lipid Res.* 2011; 52:57-67.

[39] Barter, PJ; Caulfield, M; Eriksson, M; Grundy, SM; Kastelein, JJ; Komajda, M; Lopez-Sendon, J; Mosca, L; Tardif, JC; Waters, DD; Shear, CL; Revkin, JH; Buhr, KA; Fisher, MR; Tall, AR; Brewer, B. Effects of torcetrapib in patients at high risk for coronary events. *N. Engl. J. Med.* 2007 Nov 22; 357(21):2109-2122.

[40] Badellino, KO; Wolfe, ML; Reilly, MP; Rader, DJ. Endothelial Lipase Concentrations Are Increased in Metabolic Syndrome and Associated with Coronary Atherosclerosis. *PLoS Med.* 2006 Feb; 3(2):e22.

[41] Paradis, M-E; Badellino, KO; Rader, DJ; Tchernof, A; Richard, C; Luu-The, V; Deshaies, Y; Bergeron, J; Wiedad, R; Patrick Couture, A; Bergeron, N; Lamarche, B. Visceral Adiposity and Endothelial Lipase. *J Clin Endocrinol Metab* 2006; 91:3538-3543.

[42] Kratky, D; Zimmermann, R; Wagner, EM; Strauss, JG; Jin, W; Kostner, GM; Haemmerle, G; Rader, DJ; Zechner, R. Endothelial lipase provides an alternative pathway for FFA uptake in lipoprotein lipase-deficient mouse adipose tissue. *J Clin Invest.* 2005 Jan; 115(1):161-7.

[43] Kojima, Y; Ishida, T; Sun, L; Yasuda, T; Toh, R; Rikitake, Y; Fukuda, A; Kume, N; Koshiyama, H; Taniguchi, A; Hirata, K. Pitavastatin decreases the expression of endothelial lipase both in vitro and in vivo. *Cardiovascular Research* 2010; 87:385-393.

[44] Shiu, S; Zhou, H; Wong, Y; Tan, K. Endothelial lipase and reverse cholesterol transport in type 2 diabetes mellitus. *Journal of Diabetes Investigation* 2010 June; 1(3):111-116.

[45] Pierart, C; Serrano, V; Rubio, L; Ebensperger, R; Foncea, YR. Glucose effect in the expression of endothelial lipase in human endothelial cells and in patients with diabetes mellitus type 2. *Nutr Hosp.* 2011; 26:916-921.

[46] Shiu S, Tan K, Huang Y, Wong Y. Type 2 diabetes mellitus and endothelial lipase. *Atherosclerosis* 2008; 198(2):441-7.

[47] Fujii, H; Fukuda, A; Tanaka, M; Kojima, Y; Ishida, T; Hirata, K; Fukagawa, M. Putative Role of Endothelial Lipase in Dialysis Patients with Hypoalbuminemia and Inflammation. *Am J Nephrol.* 2008; 28:974-981.

[48] Edmondson, AC; Brown, RJ; Kathiresan, S; Cupples, LA; Demissie, S; Manning, AK; Jensen, MK; Rimm. EB; Wang, J; Rodrigues, A; Bamba, V; Khetarpal, SA; Wolfe, ML; Derohnnessian, S; Li, M; Reilly, MP; Aberle, J; Evans, D; Hegele, RA; Rader DJ. Loss-of-function variants in endothelial lipase are a cause of elevated HDL cholesterol in humans. *J Clin Invest*. 2009 Apr; 119(4):1042-50.

[49] Ma, K; Cilingiroglu, M; Otvos, JD; Ballantyne, CM; Marian, AJ; Chan, AL.. Endothelial lipase is a major genetic determinant for high-density lipoprotein concentration, structure, and metabolism. *Proc. Natl. Acad. Sci. USA*. 2003; 100:2748-2753.

[50] Tang, N-P; Wang, L-S; Yang, L; Zhou, B; Gu, H-J; Sun, Q-M; Cong, R-H; Zhu, H-J; Wang, B. Protective effect of an endothelial lipase gene variant on coronary artery disease in a Chinese population. *J. Lipid Res*. 2008; 49:369-375.

[51] Jensen, MK; Rimm, EB; Mukamal, KJ; Edmondson, AC; Rader, DJ; Vogel, U; Tjønneland, A; Sørensen, T; Schmidt, EB; Overvad, K. The T111I variant in the endothelial lipase gene and risk of coronary heart disease in three independent populations. *European Heart Journal* 2009; 30:1584-1589.

[52] Shimizu, M; Kanazawa, K; Hirata, K; Ishida, T; Hiraoka, E; Matsuda, E; Iwai, C; Miyamoto, Y; Hashimoto, M; Kajiya, T; Akita, H; Yokoyama, M. Endothelial lipase gene polymorphism is associated with acute myocardial infarction, independently of high-density lipoproteincholesterol levels. *Circ. J*. 2007 June; 71(6):842-846.

[53] Durlach, V; Durlach, A; Movesayan, I; Socquard, E; Zellner, C; Clavel, C; Nazeyrollas, P; Malloy, MJ; Pullinger, CR; Kane, JP; Ducasse A; Aouizerat, BE. Association of endothelial lipase Thr111Ile polymorphism with lipid metabolism and microvascular complications in type 2 diabetic patients. *Diabetes & Metabolism* 2011 February; 37(1):64-71.

[54] Keller, PM; Rust, T; Murphy, DJ; Matico, R; Trill, JJ; Krawiec, JA; Jurewicz, A; Jaye, M; Harpel, M; Thrall, S; Schwartz, B. A High-Throughput Screen for Endothelial Lipase Using HDL as Substrate. *J Biomol Screen* 2008; 13:468-475.

[55] Darrow, AL; Olson, MW; Xin, H; Burke, SL; Smith, C; Schalk-Hihi, C; Williams, R; Bayoumy, SS; Deckman, IC; Todd, MJ; Damiano, BP; Connelly, MA. A novel fluorogenic substrate for the measurement of endothelial lipase activity *J. Lipid Res*. 2011 February 1; 52 374-382.

[56] Goodman, KB; Bury, MJ; Cheung, M; Cichy-Knight, MA; Dowdell, SE; Dunn, AK; Lee, D; Lieby, JA; Moore, ML: Scherzer, DA; Sha, D; Suarez, DP; Murphy, DJ; Harpel, MR; Manas, ES; McNulty, DE; Annan, RS; Matico, RE; Schwartz, BK; Trill, JJ; Sweitzer, TD; Wang, DY; Keller, PM; Krawiec, JA; Jaye, MC. Discovery of potent, selective sulfonylfuran urea endothelial lipase inhibitors. *Bioorganic & Medical Chemistry Letters* 2009; 19(1):27-30.

[57] Kivelä, AM; Mäkinen, PI; Jyrkkänen, H-K; Mella-Aho, E; Xia, Y; Kansanen, E; Leinonen, H; Verma, IM; Ylä-Herttuala, S; Levone, A-L. Sulforaphane inhibits endothelial lipase expression through NF-κB in endothelial cells. *Atherosclerosis* 2010; 213(1):122-128.

[58] O'Connell, D. Design and synthesis of inhibitors of endothelial lipase, a new target for raising HDL. Sackler School of Graduate Biomedical Sciences (Tufts University), 2008, available in http://proquest.umi.com/pqdlink?did=1650505691&Fmt=7&clientId=79356&RQT=309&VName=PQD

[59] Dobie, KW. Modulation of endothelial lipase expression, 2009, available in www.faqs.org/patents/app/20090156538.

[60] Qiu, G; Hill, JS. Atorvastatin decreases lipoprotein lipase and endothelial lipase expression in human THP-1 macrophages *J. Lipid Res*. 2007;.48:2112-2122.

[61] Kourimate, S; Le May, C; Langhi, C; Jarnoux, AL; Ouguerram, K; Zaïr, Y; Nguyen, P; Krempf, M; Cariou, B; Costet, P.. Dual Mechanisms for the Fibrate-mediated Repression of Proprotein Convertase Subtilisin/Kexin Type 9. *The journal of biological chemistry*, 2008 April; 283(15):9666-9673.

[62] Jin, W; Fuki, IV; Seidah, NG; Benjannet, S; Glick, JM; Rader, DJ. Proprotein convertases are responsible for proteolysis and inactivation of endothelial lipase. *J Biol Chem*. 2005 Nov 4; 280(44):36551-9.

[63] Orasanu, G; Ziouzenkova, O; Devchand, PR; Nehra, V; Hamdy, O; Horton, ES; Plutzky, J. The Peroxisome Proliferator-Activated Receptor-Agonist Pioglitazone Represses Inflammation in a Peroxisome Proliferator-Activated Receptor-Dependent Manner In Vitro and In Vivo in Mice. *JACC* 2008; 52(10):869-81.

In: Dyslipidemia ISBN: 978-1-62100-917-7
Editors: Z. Ačimovič, et.al, pp. 217-229

Chapter 8

THE ASSOCIATION BETWEEN DYSLIPIDEMIA AND NEUROPATHY– OWOLABI MO

Mayowa Ojo Owolabi*
Department of Medicine, University College Hospital, Ibadan, Nigeria
Regional Vice President (East, West and Central Africa),
World Federation for Neurorehabilitation

ABSTRACT

Microvascular damage is an important pathophysiologic mechanism underlying peripheral and autonomic neuropathy. Microvascular injury is mediated by various vascular risk factors, which are also components of the metabolic syndrome. There is increasing evidence for the role of dyslipidemia in the development and progression of diabetic peripheral and autonomic neuropathy.

Although small fiber neuropathy has been reported in non-diabetic patients with hypertriglyceridemia, the role of dyslipidemia in peripheral neuropathy is often examined in the context of other vascular risk factors which may confound the association. A number of long-term studies of large cohorts of patients with diabetes have been completed. In these studies over the course of seven or more years of follow-up, it is not glycemia but serum lipids, hypertension, and body mass index that are independently associated with the risk of developing diabetic neuropathy.

* Telephone Number: +234 802 077 5595+234 807 849 6775, Email address: mayowaowolabi@yahoo.com

There is emerging evidence that the aggregate effect of all the vascular risk factors may induce microvascular damage resulting in peripheral neuropathy. Oxidative stress provoked by oxidized LDL amongst other mechanisms may play a major role and constitute a therapeutic target. Statins and fibrates may play a role in the prevention and management of diabetic peripheral neuropathy.

INTRODUCTION

Peripheral and autonomic neuropathies are common causes of neurological disability. [1-3] Microvascular damage is an important pathophysiologic mechanism underlying these neuropathies. [2-4] Microvascular injury is mediated by various vascular risk factors, which are also components of the metabolic syndrome. [2;3;5;6] There is increasing evidence for the role of dyslipidemia in the development and progression of diabetic peripheral and autonomic neuropathy. [2;3;5;6] Although small fiber neuropathy has been reported in non-diabetic patients with isolated hypertriglyceridemia, the role of dyslipidemia in peripheral neuropathy is often examined in the context of other vascular risk factors which may confound the association. [5-10].

A number of long-term studies of large cohorts of patients with diabetes have been completed. In these studies over the course of seven or more years of follow-up, it is not glycemia but serum lipids, hypertension, and body mass index that are independently associated with the risk of developing diabetic neuropathy. [5-10] There is emerging evidence that the aggregate effect of all the vascular risk factors may induce microvascular damage resulting in peripheral neuropathy [2;3]. In this review, the association between various types of dyslipidemia and neuropathy is examined.

ROLE OF LIPID DISORDERS IN DIABETIC PERIPHERAL AND AUTONOMIC NEUROPATHIES

Diabetic peripheral neuropathy (DPN) is the most common late complication of diabetes mellitus (DM), and some of its consequences include pain, falls, foot ulceration, amputation, sleep impairment, depression and impaired quality of life. [2;3] Unfortunately, the pathogenesis of DPN is poorly understood and there is no effective treatment. [2;3] Recent studies have shown that chronic hyperglycemia measured by HbA_{1c} is neither a consistent marker of DPN occurrence nor its severity. [2;3] For instance, although the Diabetes Control Complications Trial (DCCT) reported a significant reduction in neuropathy in the intensively treated groups (with lower HbA_{1c}), the Veterans Affairs Diabetes Trial, observed a non-significant increase in autonomic neuropathy in the intensive-therapy group (with lower HbA_{1c})! [2;3] In the Veterans Affairs Diabetes Trial, and a recent Swedish study, HbA_{1c} had no significant effect on DPN. [2;3] Also in the Veterans Affairs Cooperative Study in Type 2 DM study, a 2.07% reduction in HbA_{1c} had no significant impact on diabetic peripheral or autonomic neuropathy. [2;3]

Recent studies have shown that vascular risk factors apart from hyperglycemia are involved in the development and progression of diabetic neuropathies.(2;3) The pathogenesis of DPN is complex and multi-factorial. Owolabi et al explored the relationship between total cardiovascular risk load (TCRL) using the United Kingdom Prospective Diabetes Study (UKPDS) cardiovascular risk engines, and the presence of DPN. In their cohort of patients with mean HbA1c of 6.9%, TCRL was a stronger statistical correlate and predictor of DPN than HbA_{1c}. This may have implications for prevention and monitoring of DPN particularly in those with better glycemic control. (2)

In a further study of those with DPN, Owolabi et al demonstrated that TCRL had the strongest significant correlation to DPN severity ($p =< 0.001$, rho = 0.285) with age and dyslipidemia (LDL/HDL ratio), which are also components of the TCRL, emerging as independent statistical predictors of DPN severity in multivariate analysis [2;3].

Pathophysiology of Diabetic Neuropathy: Beyond Hyperglycemia

Until recently, it was thought that hyperglycemia was the driving force underlying the development of diabetic neuropathy. This was based originally on results from the DCCT Diabetes Control and Complications Trial (DCCT). In the DCCT, type 1 diabetic subjects receiving intensive therapy with an average glycosylated hemoglobin (HbA1c) of 7.2% had a reduced 60% cumulative incidence of diabetic neuropathy when compared to patients receiving conventional treatment (average HbA1c of 9.0%). [10]

However, the continuing longitudinal study of the DCCT, the Epidemiology of Diabetes Complications and Interventions Cohort (EDIC) yielded unanticipated results 20 years later. Within a year of discontinuing the DCCT and beginning EDIC, the glycemic control in the two treatment groups equalized to an average HbA1c of 8%. All patients were examined annually for diabetic neuropathy; one decade later, patients from the intensive-DCCT cohort had a lower incidence of diabetic neuropathy compared to patients from the conventional-DCCT cohort, despite 10 years of convergent glycemic control. Although the underlying mechanism(s) of this result was not determined, the cohort with a higher incidence of DPN had more dyslipidemia than the conventional cohort. [10]

This interesting, unanticipated finding is further supported by the Eurodiab Trial, a longitudinal study of over 3,000 individuals with type 1 diabetes [5;6;11]. Of the 1,200 subjects who did not have diabetic neuropathy at baseline, hypertension, serum lipids and body mass index were each independently associated with the risk of developing diabetic neuropathy during a 7-year follow-up period. Of these risk factors, dyslipidemia was closely linked with the onset and progression of diabetic neuropathy. [10]

In support of these findings, Wiggins et al evaluated the mechanisms underlying diabetic neuropathy progression using indices of sural nerve morphometry obtained from two identical randomized, placebo-controlled clinical trials. They found elevated triglycerides to be the only clinical parameter that correlated with a loss of myelinated fiber density, independent of disease duration, age, diabetes control, or other variables [12].

Dyslipidemia and Neuropathy: The Emerging Evidence

The emerging idea that dyslipidemia contributes to the development of diabetic neuropathy may explain the earlier incidence of diabetic neuropathy in individuals with type 2 compared to type 1 diabetes. Lipid profiles are commonly abnormal early in the course of type 2 diabetes in a temporal pattern that correlates with the presence of diabetic neuropathy [10]. In contrast, lipid profiles are nearly always normal in type 1 patients at the time of diabetes diagnosis [8]. Dyslipidemia develops later in the course of type 1 diabetes, and these abnormal lipid profiles coincide with the delayed onset and progression of diabetic neuropathy. [10]

Accumulating data from several large scale trials of patients with type 2 diabetes also point to early dyslipidemia as a major independent risk factor for the development of diabetic neuropathy [10]. In the United Kingdom Prospective Diabetes Study (UKPDS), newly diagnosed type 2 patients were randomized into either intensive treatment with an oral hypoglycemic agent or insulin or conventional treatment with diet. After 10 years, intensive treatment resulted in approximately 1% lower HbA1c vs. conventional treatment but there was no difference in the development of diabetic neuropathy between the two groups, which had similar lipid and blood pressure [10].

This finding, at first unexpected in light of the earlier DCCT data, was supported by the Veterans Affairs Cooperative Study (VACS), which demonstrated no difference in the prevalence of diabetic neuropathy in type 2 diabetic patients over a 2-year period comparing standard and intensive glycemic control [10]. These results suggested that independent factors other than glycemic control are critical to the development of diabetic neuropathy [8].

As with EDIC and Eurodiab, analysis of the UKPDS and VACS data points to dyslipidemia as a critical independent factor for the development of diabetic neuropathy [8]. Type 2 diabetes clusters with risk factors for coronary heart disease including obesity, hypertension, and dyslipidemia; individuals with two or more of these factors are diagnosed with the metabolic syndrome. In a cross-sectional study of type 2 diabetic subjects, those with metabolic syndrome were twice as likely to have diabetic neuropathy [13], and the driving factor was dyslipidemia. This supports the observation by Owolabi et al where aggregated vascular risk factors for stroke or coronary heart disease was the strongest correlate of the presence as well as the severity of DPN. [2;3]

In a European study of 85 type 2 diabetic patients with at least two additional metabolic syndrome parameters, the prevalence of microvascular complications, including diabetic neuropathy, increased with each additional parameter present(13); abnormalities in (low-density lipoprotein) LDL profiles were more closely related to diabetic neuropathy than hyperglycemia. [2;3;10]

Finally, prospective studies of patients with idiopathic neuropathy, confirm a higher prevalence of hyperlipidemia than impaired glucose tolerance or hypertension, suggesting that dyslipidemia is an essential factor underlying nerve injury. Collectively, this evolving and exciting literature links dyslipidemia to the development and progression of diabetic neuropathy. [2;3;10]

Neuropathy may develop as a result of endothelial dysfunction, which can cause a reduction in neuronal blood flow. Endoneural capillary density is reduced and correlates with decreased density of myelinated fibres in neuropathic diabetic patients. [2;3;10;14]. These effects may impair the function of motor, sensory and autonomic neurons in patients with diabetes.

[2;3;8;14] Endothelial injury resulting from multiple vascular risk factors particularly dyslipidemia seem to pay a major role. [2;3;8;14]

Indeed, intensive treatment of the cardiovascular risk factors associated with diabetes, including hyperglycemia, dyslipidemia and hypertension, has been shown to reduce the risk of microvascular complications including diabetic neuropathy [2;3;8;14].

Prospective studies tended to show a positive association between lipid levels and diabetic microangiopathy but results are inconsistent with regard to which lipid fraction is most predictive. [8] Whereas hypertrygliceridemia is an independent risk factor for neuropathy, evidence is mounting on the possible role of modified LDL-cholesterol in DPN.

The Role of Modified Lipids in the Pathophysiology of Diabetic Neuropathy

Using cell culture and mouse models of diabetic neuropathy, it has been suggested that oxidized LDLs (oxLDLs) are one notable 'lipid factor' responsible for nervous system injury (10) oxLDLs are critically involved in endothelial cell dysfunction, evident from the large body of literature implicating oxLDLs in atherosclerotic lesion formation [10] oxLDL is strongly cytotoxic, which may explain the areas of necrosis detected within atherosclerotic lesions [10]. In man, oxLDL is a highly analytic marker for macrovascular disease, including stroke and myocardial infarction [10]. In patients with types 1 and 2 diabetes, serum levels of oxLDL in proportion to total LDL particles are associated with diabetic neuropathy. [10] oxLDLs cause apoptotic injury and death in both endothelial cells and neurons [10]. In endothelial cells, oxLDLs induce multiple events associated with apoptotic injury, including Bid degradation, cytochrome c release, and caspase-3 activation [15;16] In neurons, oxLDLs induce DNA fragmentation, increase reactive oxygen species (ROS) and activate a caspase-3-dependent death mechanism [15;16]. These data suggest that, in diabetes, neurons are exposed to both glucose and oxLDL whichindependently increase ROS which in conjunction with hypertension inflicts endothelial and neuronal injury (Figure 1). [8;10;15;16]

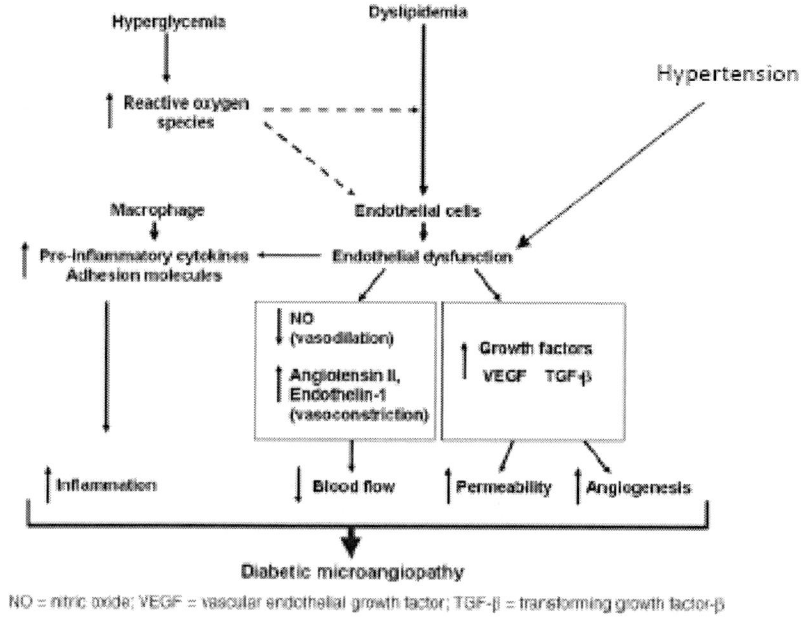

Figure 1. Mechanism whereby hyperglycemia and dyslipidemia can produce microvascular complications. Modified from Leiter L Diabetes Research and Clinical Practice 68S2 (2005) S3–S14 [8]. (With permission from Elsevier)

Hypertriglyceridemia and Peripheral Neuropathy

Apart from oxidized LDL, hypertriglyceridemiamay also be important in the development of neuropathy. The relationship between hypertriglyceridemia and neuropathy has been explored in several case series. Mc Manis in 1994 described a group of 6 patients with moderately elevated cholesterol levels and exceedingly high serum triglyceride levels. Each had mild, slowly progressive neuropathy characterized by pain in the feet, without proximal extension or involvement of the hands. [17] Weakness and autonomic symptoms and signs were absent. No other cause of peripheral neuropathy was found. In one patient, symptoms resolved with correction of hypertriglyceridemia. [17] Despite the possible confounding effect of elevated cholesterol, this observation suggests that marked increases in serum triglycerides may cause length-dependent painful small-fiber neuropathy.

In a prospective study, Drory et al evaluated 16 patients with marked hypertriglyceridemia without other causes of neuropathy, using nerve conduction and autonomic function tests. [18] Six subjects (37%) showed mild signs of asymptomatic polyneuropathy. They concluded that hypertriglyceridemia may be associated with a mild subclinical axonal polyneuropathy. [18]

This association was further explored in a larger study by [7] Kassem et al in 2005. They investigated 24 patients with a high triglyceride level (>300mg/dl) without neurological complaints. They observed that over 2/3 of the patients had significant delays in the sural and median nerve distal latencies. In addition, over half of the patients had a significant decrease in the motor nerve conduction velocities while about 1/3 had a decrease in the sensory nerve conduction velocities. The amplitudes were relatively spared. This pattern does not fulfil the criteria for small fibre or axonal neuropathy.

Nevertheless, collectively these studies suggest that severe hypertriglyceridemia is associated with peripheral neuropathy. This conclusion is further supported by a recent case report [19]. in which a patient with extremely high serum triglyceride (> 10,000 mg/dl) presented with severe peripheral neuropathy. However larger studies are warranted to fully characterize the clinical and neurophysiological presentation of hypertriglyceridemia-induced neuropathy.

Role of Lipid Disorders in HIV-Associated Neuropathies

In HIV populations that are aging due to improved longevity with combination antiretroviral therapy, both hypertriglyceridemia and sensory neuropathy have become increasingly common. Banerjee et al explored the relationship between hypertriglyceridemia and sensory neuropathy using a comparative study design in 436 HIV patients. [20] They observed that among HIV-positive patients, those with triglyceride levels ≥ 244 mg/dl were more likely to have sensory neuropathy than those with levels ≤ 142 mg/dl independent of other known risk factors (adjusted odds ratio 2.7, 95% confidence interval 1.4-5.5). [20] This observation further strengthens the link between hypertriglyceridemia and peripheral neuropathy.

Abetalipoproteinemia and Neuropathy

Rare disorder of lipid metabolism may also be associated with neuropathy. Abetalipoproteinemia, or Bassen-Kornzweig syndrome, is a rare autosomal recessive disorder with impaired absorption of fat and fat-soluble vitamins from food. It is caused by a deficiency of apolipoprotein B-48 and B-100, which are used in the synthesis and exportation of chylomicrons and VLDL respectively. [21] The association between abetalipoproteinemia and neuropathy has been explored only in few case reports.

In a case series of three sisters with abetalipoproteinemia, clinically, a progressive sensory neuropathy was found. [21] There was a diminution in the amplitude of sensory action potentials and a slight-to-moderate slowing in maximum sensory conduction velocity, initially most marked in distal portions of the nerves. Motor conduction was normal, although EMG indicated subclinical signs of partial chronic denervation. The sural nerves showed a decreased number of large fibers and evidence of paranodal demyelination and regeneration of unmyelinated fibers was found. [21] This pattern suggests a sensorimotor neuropathy.

Tangier's Disease and Neuropathy

Tangiers disease is also known as familial alpha lipoprotein deficiency (analphalipoproteinemia). There are a few case reports describing the presence of chronic neuropathy in patients with this disease. This neuropathy is characterised by lipid accumulation in and subsequent damage of Schwann cells of myelinated and unmyelinated fibres. [22;23] Rarely, there is suggestion of damage to the vasa nervorum with progressive sensorimotor distal neuropathy [24].

Dyslipidemia and Autonomic Neuropathies

The role of isolated dyslipidemia in autonomic neuropathies is not well characterized. However, as highlighted above, several studies have shown dyslipidemia to be an independent predictor of autonomic neuropathy in diabetes mellitus. Indeed Voulgari et al reported an odds ratio of 1.35 for developing cardiac autonomic neuropathy in those with elevated LDL-cholesterol, and an odds ratio of 1.30 in those with elevated triglycerides. However these findings require further confirmation in prospective large scale studies. [25]

Statins, Fibrates and Neuropathy

If the idea that dyslipidemia contributes to the development of diabetic peripheral and autonomic neuropathy is true, lipid lowering drugs may be beneficial in the treatment of diabetic neuropathy.

Fenofibrate is a Peroxisome Proliferator-Activated Receptor-alpha PPARα agonist that lowers plasma lipids by improving their removal by the liver and improving fatty acid metabolism. In genetic dyslipidemia in mice, including ApoE knockout, leptin deficient, and LDL receptor knockout mice, fenofibrate improved the lipid profile and increased high-density lipoprotein. These lipid improvements correlate with prevention of insulin resistance and atherosclerosis. The Fremantle Diabetes Study was an observational investigation of 1,237 patients with type 2 diabetes. The data suggest that therapy with a statin or fibrate protects against diabetic peripheral sensory neuropathy, but calls for confirmatory evidence via a randomized clinical trial. [10]

Clinical evidence supporting the use of statins to prevent neuropathy is limited, although studies in animals have shown promising results. Studies in diabetic rats showed that rosuvastatin treatment improves nerve conduction velocity (NCV) and restores perception to thermal pain. Recent preclinical studies with rosuvastatin have also demonstrated improvements in NCV in a streptozotocin-induced rat model of diabetes, as well as improvements in hyperalgesia . [10]

The clinical benefits of statins on erectile dysfunction (ED) have not been established. There are some data to suggest that statins may improve nocturnal penile tumescence. Preclinical studies with rosuvastatin would tend to support a beneficial effect in ED. Studies are required to confirm whether this benefit of statin treatment is translated to diabetic patients. [8]

On the flipside of the coin, there is some evidence that statins may induce neuropathy. Lo described three patients who developed small fibre neuropathy after 1 month of statin therapy with clinical resolution upon prompt drug withdrawal.(26) All patients showed abnormal sympathetic skin responses (SSR) in comparison with controls. SSRs returned to normal in tandem with clinical improvement. One patient redeveloped small and large fibre neuropathy when the similar drug was re-administered. [26]

Therefore more properly designed long- term randomised controlled clinical trials with exhaustive risk-benefit analysis are warranted to determine the role of statins, fibrates and ezetimibe in the treatment and prevention of neuropathies.

CONCLUSION

While hyperglycemia may play a key role in diabetic neuropathies in poorly controlled diabetes, dyslipidemia and other vascular risk factors become more important with improved glycemic control. [2;3;10] Therefore, aggregated vascular risk load assessment is the preferred integrated marker for diabetic neuropathy. This needs to be further confirmed in prospective studies. [2;3] There is inconclusive evidence regarding the type of lipid disorder with the strongest association to diabetic neuropathy. This needs to be explored in large scale multicentre prospective studies.

The specific clinical and neurophysiological patterns of the different lipid disorders associated with neuropathy require better characterization. Furthermore, the underlying molecular mechanisms need to be well studied and outlined for further drug refinement and subcellular targeting [2;3].

REFERENCES

[1] Owolabi MO, Bower JH, Ogunniyi A. Mapping Africa's way into prominence in the field of neurology. *Arch. Neurol.* 2007 Dec;64(12):1696-700.

[2] Owolabi MO, Ipadeola A, Adeleye JO. Aggregate cardiovascular risk is a stronger statistical correlate of clinically evident diabetic peripheral neuropathy than HbA1c alone. *J. Natl. Med. Assoc.* 2010 Aug;102(8):707-12.

[3] Owolabi MO, Ipadeola A. Total vascular risk as a stronger correlate of severity of Diabetic Peripheral Neuropathy in Nigerian Africans. *Ethnicity and diseases* 2011;In press.

[4] Dyck PJ, Davies JL, Clark VM, Litchy WJ, Dyck PJ, Klein CJ, et al. Modeling chronic glycemic exposure variables as correlates and

predictors of microvascular complications of diabetes. *Diabetes Care* 2006 Oct;29(10):2282-8.

[5] Tesfaye S, Malik R, Ward JD. Vascular factors in diabetic neuropathy. *Diabetologia 1*994 Sep;37(9):847-54.

[6] Tesfaye S, Chaturvedi N, Eaton SE, Ward JD, Manes C, Ionescu-Tirgoviste C, et al. Vascular risk factors and diabetic neuropathy. *N. Engl. J. Med.* 2005 Jan 27;352(4):341-50.

[7] Kassem HS, Azar ST, Zantout MS, Sawaya RA. Hypertriglyceridemia and peripheral neuropathy in neurologically asymptomatic patients. *Neuro Endocrinol. Lett* 2005 Dec;26(6):775-9.

[8] Leiter LA. The prevention of diabetic microvascular complications of diabetes: is there a role for lipid lowering? *Diabetes Res. Clin. Pract.* 2005 Jun;68 Suppl 2:S3-14.

[9] Tesfaye S, Tandan R, Bastyr EJ, III, Kles KA, Skljarevski V, Price KL. Factors that impact symptomatic diabetic peripheral neuropathy in placebo-administered patients from two 1-year clinical trials. *Diabetes Care* 2007 Oct;30(10):2626-32.

[10] Vincent AM, Hinder LM, Pop-Busui R, Feldman EL. Hyperlipidemia: a new therapeutic target for diabetic neuropathy. *J. Peripher. Nerv. Syst.* 2009 Dec;14(4):257-67.

[11] Tesfaye S. Advances in the management of diabetic peripheral neuropathy. *Curr. Opin. Support. Palliat. Care* 2009 Jun;3(2):136-43.

[12] Wiggin TD, Sullivan KA, Pop-Busui R, Amato A, Sima AA, Feldman EL. Elevated triglycerides correlate with progression of diabetic neuropathy. *Diabetes* 2009 Jul;58(7):1634-40.

[13] Isomaa B, Henricsson M, Almgren P, Tuomi T, Taskinen MR, Groop L. The metabolic syndrome influences the risk of chronic complications in patients with type II diabetes. *Diabetologia* 2001 Sep;44(9):1148-54.

[14] Dyck PJ, Giannini C. Pathologic alterations in the diabetic neuropathies of humans: a review. *J. Neuropathol. Exp. Neurol.* 1996 Dec;55(12):1181-93.

[15] Keller JN, Hanni KB, Kindy MS. Oxidized high-density lipoprotein induces neuron death. *Exp. Neurol.* 2000 Feb;161(2):621-30.

[16] Vindis C, Elbaz M, Escargueil-Blanc I, Auge N, Heniquez A, Thiers JC, et al. Two distinct calcium-dependent mitochondrial pathways are involved in oxidized LDL-induced apoptosis. *Arterioscler. Thromb. Vasc. Biol.* 2005 Mar;25(3):639-45.

[17] McManis PG, Windebank AJ, Kiziltan M. Neuropathy associated with hyperlipidemia. *Neurology* 1994 Nov;44(11):2185-6.

[18] Drory VE, Groozman GB, Rubinstein A, Korczyn AD. Hypertriglyceridemia may cause a subclinical peripheral neuropathy. *Electromyogr. Clin. Neurophysiol.* 1999 Jan;39(1):39-41.

[19] Hou R, Goldberg AC, Tobin GS. A case of severe neuropathy associated with hypertriglyceridemia. Endocr Pract 2008 Nov;14(8):1020-2.

[20] Banerjee S, McCutchan JA, Ances BM, Deutsch R, Riggs PK, Way L, et al. Hypertriglyceridemia in combination antiretroviral-treated HIV-positive individuals: potential impact on HIV sensory polyneuropathy. *AIDS* 2011 Jan 14;25(2):F1-F6.

[21] Wichman A, Buchthal F, Pezeshkpour GH, Gregg RE. Peripheral neuropathy in abetalipoproteinemia. *Neurology* 1985 Sep;35(9):1279-89.

[22] Hager H, Zimmermann P. [Peripheral nerve changes in Tangier disease. Comparative light-, electron-microscopic and cytometric studies (author's transl)]. *Acta. Neuropathol.* 1979 Jan 12;45(1):53-9.

[23] Hager H, Zimmermann P. Morphology and cytometry of sural nerve in Tangier disease. *Neuropatol. Pol.* 1982;20(1-2):123-8.

[24] Marbini A, Gemignani F, Ferrarini G, Maccari S, Lucci B, Bragaglia MM, et al. Tangier disease. A case with sensorimotor distal polyneuropathy and lipid accumulation in striated muscle and vasa nervorum. *Acta. Neuropathol.* 1985;67(1-2):121-7.

[25] Voulgari C, Psallas M, Kokkinos A, Argiana V, Katsilambros N, Tentolouris N. The association between cardiac autonomic neuropathy with metabolic and other factors in subjects with type 1 and type 2 diabetes. *J. Diabetes Complications* 2011 May;25(3):159-67.

[26] Lo YL, Leoh TH, Loh LM, Tan CE. Statin therapy and small fibre neuropathy: a serial electrophysiological study. *J. Neurol. Sci.* 2003 Apr 15;208(1-2):105-8.

In: Dyslipidemia
Editors: Z. Ačimovič et.al, pp. 231-245
ISBN: 978-1-62100-917-7

Chapter 9

FENOFIBRATE-MEDIATED ANTI-ATHEROGENIC ACTIVITY: LONG-TERM FOLLOW-UP STUDY OF RHEUMATOID ARTHRITIS PATIENTS BY COMPARING WITH STATIN THERAPY

Makoto Goto
Division of Anti-Ageing and Longevity Sciences,
Faculty of Medical Technology,
Department of BioMedical Engineering, Toin University of Yokohama
Aoba-Ku, Yokohama 225-8502, Japan

ABSTRACT

Objectives

Fenofibrate exerts multiple pleiotropic effects as a selective peroxisome proliferator-activated receptor alpha (PPAR-α) agonist in addition to its well-characterized lipid-lowering effects.

Fenofibrate suppressed the production of pro-inflammatory cytokines through the inhibition of NFκB pathway and ubiquitin-proteasome system, leading to the decrease in the incidence of atherosclerosis and type II diabetes mellitus (DM) that are tightly associated with chronic inflammation.

Anti-inflammatory and anti-oxidant activities of fenofibrate have been studied in the chronic inflammatory diseases such as rheumatoid arthritis (RA) with a higher risk of atherosclerosis-related death.

Methods

In this chapter, our long-term follow-up study on the anti-inflammatory and anti-oxidative activities of fenofibrate in RA were compared with the parallel study of statins therapy and control concerning to the changes in the serum levels of lipid profile and inflammatory parameters including C-reactive protein (CRP), erythrocyte sedimentation rate (ESR), matrix metalloproteinase-3 (MMP-3), pentosidine and homocysteine.

Results

Both serum lipid profile and the most inflammatory parameters such as CRP, ESR and MMP-3 remained basically stable in all the 3 groups during the 18 months study. Although serum pentosidine and homocysteine levels elevated significantly in the control group without lipid-lowering agents, the two parametersdid not change significantly in fenofibrate and statins groups. Prednisolone dosage decreased significantly in all the 3 groups.

Conclusions

Lipid-lowering agents such as fenofibrate and statins may play a role as anti-atherogenic and anti-oxidant agents in the chronic inflammatory disease like RA.

ABBREVIATIONS

AGE: advanced glycation end- product,
CRP: C-reactive protein,
DAS: diseaseactivity score,
DM: diabtes mellitus,
DMARD: disease-modifying anti-rheumatic drug,

ELISA:	enzyme-linked sandwich assay,
ESR:	erythrocyte sedimentation rate,
HDL:	high density lipoprotein,
HPLC:	high performance liquid chromatography,
LDL:	low density lipoprotein,
MMP-3:	matrix metalloproteinase 3,
MTX:	methotrexate,
PPAR:	peroxisome proliferator-activated receptor,
OA:	osteoarthritis,
PSL:	prednisolone,
RA:	rheumatoid arthritis,
Sjs:	Sjogren syndrome,
SLE:	systemic lupus erythematosus,
TARA:	trial of atrovastatin in rheumatoid arthritis,
TC:	total cholesterol,
TG:	triglyceride.

INTRODUCTION

The average life-span of the patients suffering from a variety of rheumatic diseases such as rheumatoid arthritis (RA) and systemic lupus erythematosus (SLE) has been decreased compared with the general population despite of the rapid improvement of disease control [1, 2, 3, 4, 5]. The major causes of death in the rheumatic diseases include infections of various kinds that are presumably associated with accelerated aging resulted from immunological abnormalities and immunosuppressive therapy inherent to the diseases [6]. These are followed by cardiovascular accidents associated with atherosclerosis-related diseases such as myocardial and cerebral infarctions [7, 8, 9].

Dyslipidemia, a risk factor for atherosclerosis, has been suspected to be an adverse event of steroid therapy in RA and SLE [10, 11, 12]. However, the anti-atherogenicacitivity of steroid has been described [13, 14]. Other risk factors for atherosclerosis include homocysteine and pentosidine: a family of advanced glycation end products (AGEs) generated under oxidative stress and aging [9, 15, 16, 17, 18, 19, 20].

The Trial of Atorvastatin in Rheumatoid Arthritis (TARA Study), comparing atorvastatin, a lipid-lowering statin, to a placebo was conducted by McCarey et al. in 2004. Six-month administration of atorvastatin significantly improved the Disease Activity Score 28 (DAS28), C-reactive protein (CRP), erythrocyte sedimentation rate (ESR) and number of swollen joints. The anti-inflammatory effects of atorvastatin thus received special attention [21]. A report in Nature also detailed the clinical anti-inflammatory activity of fenofibrate, a lipid-lowering fibrate agonist of peroxisome proliferator-activated receptor α (PPAR-α) [22]. We thus anticipated that lipid-lowering agents with pleiotropic effects may serve also as anti-inflammatory agents in RA.

We compared the lipid-lowering activity and anti-inflammatory activity of fenofibrate and statins administered in the patients with RA complicated with dyslipidemia. In the previous communication,fenofibrate showed more anti-inflammatory and anti-dyslipidemic activities than stainsduring a relatively short period [23].Thus, we have planned to study the long term effects on the other risk factors for atherosclerosis: serum homocysteine and pentosidine inthe RA with dyslipidemia treated by fenofibrate or statins.

Table 1. Patients' background

	Control	Fenofibrate	Statins
No. of patients (male/female)	48 (17/31)	38 (8/30)	19 (6/13)
Age (years)	57.8±15.4	64.0±9.1*	63.2±12.2
Complications	4 (Sjs=1,OA=2,Gout=1)	7 (Sjs=4,OA=2,Gout=1)	2 (Sjs=1,OA=1)
CRP (mg/dL)	0.87±1.55	1.01±1.63	0.40±0.49
MMP-3 (ng/mL)	199.0±203.9	210.1±236.4	208.8±249.0
ESR (mm/h)	28.7±19.9	38.3±32.5	36.5±14.5*
No. of patients receiving concomitant therapy (%)			
MTX	21 (43.8%)	10 (26.3%)	7 (36.8%)
Other DMARDs	31 (64.6%)	31 (65.8%)	12 (63.2%)
Biologics	9 (18.8%)	3 (7.9%)	0 (0.0%)*
Steroid	32 (66.7%)	23 (60.5%)	9 (47.4%)

Mean±SD.

* $p<0.05$ significant compared with control by Fisher's exact test.

CRP C-reactive protein, MMP-3 matrix metalloproteinase 3, ESR erythrocyte sedimentation rate, SjsSjogren syndrome, OA osteoarthritis, MTX methotrexate, DMARD disease modifying antirheumatic drug.

MATERIALS AND METHODS

A total of 105 RA patients complicated with dyslipidemia were randomly assigned to the fenofibrate group (Lipidil®, Kaken Pharmaceutical Co., Ltd., 200 mg/day, n = 38), the statins group (n = 19) and control group without lipid-lowering agents (n=48). Statins used included atorvastatin (Lipitor:n=16), pitavastatin (Livalo:n=2) and pravastatin (Mevalotin:n=1). Diagnosis of dyslipidemia was made if one of the following criteria was met: total cholesterol (TC) ≥ 220 mg/dL, low-density lipoprotein-cholesterol (LDL-C) ≥ 140 mg/dL, high-density lipoprotein-cholesterol (HDL-C) < 40 mg/dL or triglycerides (TG) ≥ 150 mg/dL, as previously described [23].

The baseline blood chemistry (serum lipids profile, homocysteine, and pentosidine) was compared with those obtained 18 months after the start of the study. Homocysteine (nmol/ml) was examined by HPLC [24] and pentosidine (ug/ml) was assayed by ELISA as previously described [25].

The effects on RA activity were assessed by comparing the serum levels of CRP, ESR, MMP-3 and the dosage of prednisolone (PSL). An informed consent form, which had been formally approved by the Ethical Committee of Toin University of Yokohama, was used to explain the purpose of this study and the drugs to be administered to the patients.

The numerical results are expressed as mean ± S.D. For the statistical analysis, a Fisher's exact test was used to assess the patients' background, Wilcoxon signed rank test for the comparison between the baseline and post-treatment data, Mann-Whitney U test for intergroup comparison. Values below $p < 0.05$ were regarded as significant.

RESULTS

Patients'Background

Patients' data are shown in Table 1 and Figure 1. Each group included a few patients complicated with Sjogren syndrome (Sjs), osteoarthritis (OA) or gout as indicated in Table 1. Mean±SD age was 64.0±9.1 years old in the fenofibrate group (M=8,F=30), 63.2±12.2 years old in the statin group (M=6, F=13) and 57.8±15.4 years old in control group (M=17, F=31). Although there were no significant differences in sex distribution among 3 groups, the age in fenofibrate group was significantly older than control group ($p<0.05$). The disease activity assessed by CRP (mg/dl), ESR (mm/h) and MMP-3 (ng/ml) was similar between 3 groups except that ESR in the statins group was significantly higher than control group (p<0.05).

Among disease-modifying anti-rheumatic drug (DMARD) users, methotrexate (MTX) was prescribed to 10 of 38 patients (26.3%) in the fenofibrate group, 7 of 19 (36.8%) in the statins group and 21 of 48 (43.8%) in the control group. In Japan, the dosage of MTX has been officially regulated below 8 mg/week during the present study. In addition, all the MTX users were supplemented with folate. No biologic users were in the statins group (control vs statins; p<0.05), but 3 of 38 (7.9%) in the fenofibrate and 9 of 48 (18.8%) in the control group. Steroid (prednisolone <10mg/day) was similarly prescribed to 32 of 48 (66.7%) in control group, 23 of 38 (60.5%) in fenofibrate and 9 of 19 (47.4%) in statins groups.

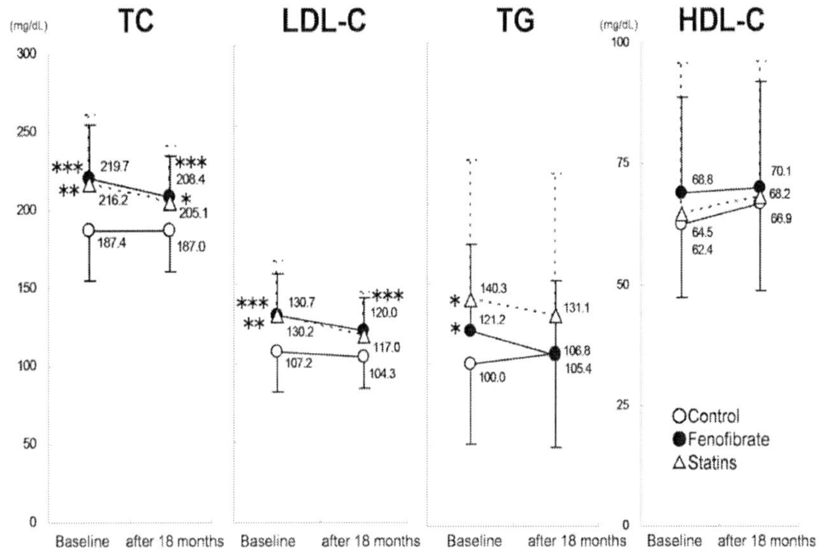

Mean±SD.Open circles represent control. Closed circles represent fenofibrate. Open tiangles represent statins.* p<0.05, ** p<0.01, *** p<0.001 significant compared with control by Wilcoxon signed rank test.TC total cholesterol, LDL-C low-density lipoprotein-cholesterol, TG triglycerides, HDL-C high-density lipoprotein-cholesterol.

Figure 1.Changes in serum lipid levels.

Baseline HDL-C was similar between 3 groups, but baseline TC, LDL-C and TG in control group were significantly lower thanfenofibrate group ($p<0.001$, $p<0.001$ and <0.05) and statins group ($p<0.01$, $p<0.01$ and <0.05). The baseline lipid profile (TC, HDL-C, LDL-C, and TG) was comparable between fenofibrate and statins groups as shown in Figure 1.

No adverse episode including changes or addition of other DMARDs was recorded during the study.

Improvement in Serum Lipid Profiles

As shown in Figure 1, lipid profile in fenofibrate group improved after 18 months (TC: from 219.7±34.6 to 208.4±25.6, HDL-C: from 68.8±20.0 to 70.1±22.1, LDL-C: from 130.7±26.3 to 120.0±21.0, and TG: from 121.2±54.0 to 105.4±47.1), but the changes did not reach a statistical significance. In both control and statins groups, the lipid profile basically remained unchanged.

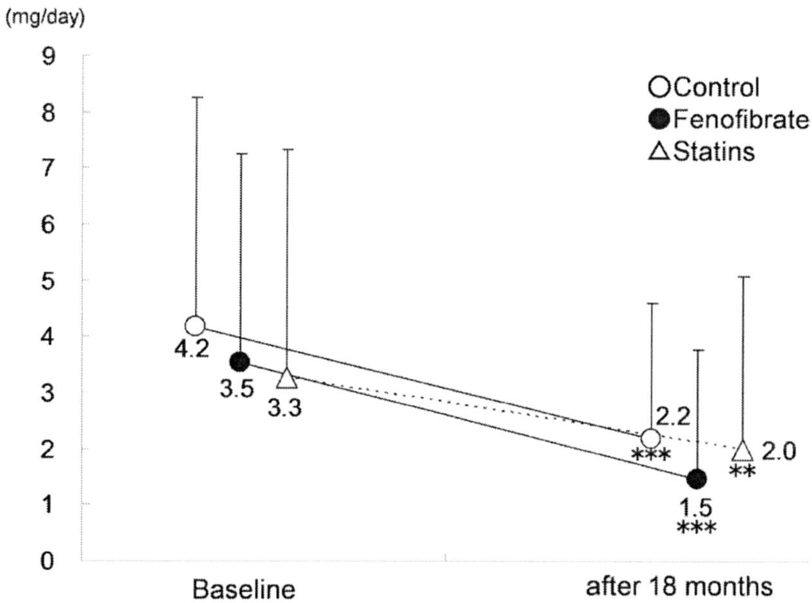

Mean±SD .Open circles represent control. Closed circles represent fenofibrate. Open tiangles represent statins.** $p<0.01$, *** $p<0.001$, Wilcoxon signed rank test versus baseline.

Figure 2.Changes in prednisolone (PSL) dosage.

Improvement of Inflammatory Activity

The dosage of PSL decreased significantly in all the three groups after the 18 months compared with the baseline (fenofibrate group <0.001,controlgroup <0.001, and statins group <0.01) (Figure 2). Among 3 groups, the reduction rate of PSL was more pronounced, but not statistically signaificant in fenofibarte group (-62.1±33.9%) than control (-47.6±48%) and statins groups (-44.4±37.5%). Although the CRP and ESR levelsinsignificantly decreased from 1.01±1.63 to 0.57±1.01 mg/dl and from 38.3±32.5 to 31.8±24.7 mm/h in the fenofibrate group, the 2 parameter levels in statins and control groups remained unchanged. MMP-3 level insignificantly decreased in all the groups.

Changes in Serum Pentosidine and Homocysteine Levels

Both pentosidine and homocysteine levels increase with normal aging and inflammation [16, 19, 20]. As indicated in Figure 3, serum Δpentosidine (97.7±198.9%, p<0.01) and Δhomocysteine (10.6±31.4%, p<0.05) increased significantly after 18 months in control group without lipid-lowering treatment. In contrast, Δpentosidine did not change significantly in fenofibrate (44.3 ±233.4%) and statins (47.4±187.6%) groups.

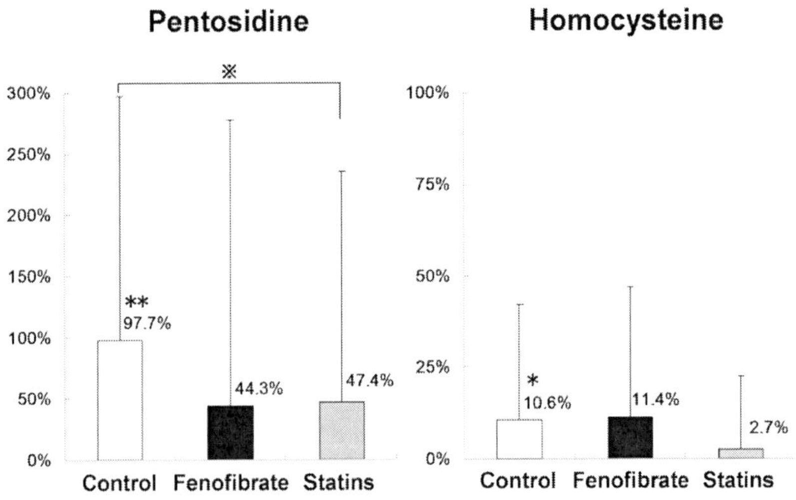

Mean±SD.* p<0.05, ** p<0.01, Wilcoxon signed rank test versus baseline. ※ p<0.05, Mann-Whitney U test versus control.

Figure 3. % changes in Δpentosidine and Δhomocysteine.

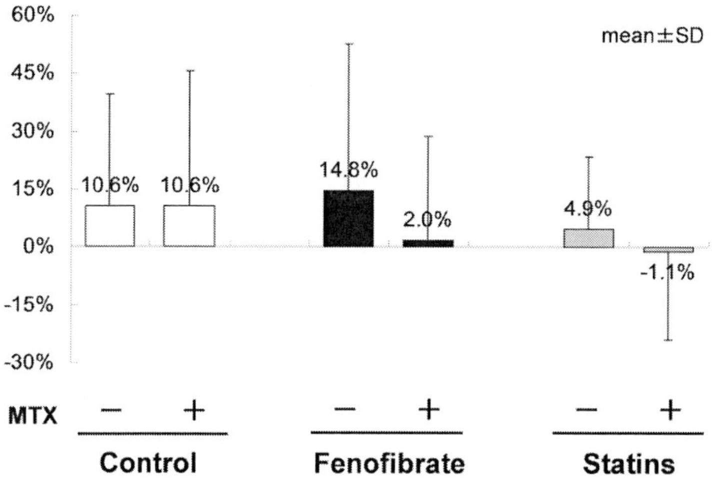

Figure 4. % changes in Δhomocysteine by MTX treatment.Mean±SDΔhomocysteine level was compared with and without MTX after 18 months study in 3 groups.

Δpentosidine in statins group decreased significantly compared with the control ($p<0.05$). Pentosidine both in fenofibrate (from 0.09±0.10 to 0.09±0.13) and statins (from 0.12±0.12 to 0.13±0.16) groups was unchanged, while pentosidine in control group (0.11±0.12 to 0.15±0.16) increased significantly compared with fenofibrate group ($p<0.05$) after 18 months treatment.

Also, Δhomocysteine remained unchanged in fenofibrate (11.4±35.5%) and statins (2.7±19.8%) groups (Figure 3). However, homocysteine level increased insignificantly in 3 groups.

Figure 4 showed the contribution of MTX to serum homocysteine level. Δhomocysteine in normal control group insignificantly elevated irrespective of MTX therapy. However, Δhomocysteine in both fenofibrate and statins groups insignificantly decreased by MTX usage.

Serum pentosidine level did not change significantly between with and without MTX in all the 3 groups(data not shown). Thus, low dose MTX did not affect the serum levels of homocysteine and pentosidine significantly.

DISCUSSION

Both elevation of serum homocysteine and pentosidine were suppressed by the use of fenofibrate or statins compared with control group without lipid-lowering agents, though the lipid-lowering agents including fenofibrate and statins insignificantly improved serum lipid levels after the 18 months treatment.

Homocysteine: a traditional risk factor for cardiovascular diseases, has been increased with normal aging and cardiovascular diseases that may be associated with chronic inflammation [26,27,28,29]. Although an increase in homocysteine level in general population with dyslipidemia by fenofibrate has been frequently reported [30, 31, 32], this may not the case in Japanese RA patients as was shown in the present study.

Serum homocysteine was reported to be elevated in RA treated with higher dose(25~40mg/week) MTX [33] or evenby low-dose MTX [34], though concomitant folate supplementation may decrease the plasma homocysteine level [34] and most anti-RA drugs (NSAIDs, steroids, biologics, leflunomide, COX-2 inhibitors) may not affect major adverse effects on cardiovascular risk factors [35]. In the present study, 3~40% patients were treated with low dose MTX in combination with folatesupplementation that may not affect homocysteine metabolism, as was shown in Figure 4.

Homocysteine may be produced by inflammatory mechanism through NFκB pathway that may be blocked through PPARα(fenofibrate) and/or statins [22,36].

Pentosidine: a member of AGEs and a biomarker of chronic inflammation has been reported to elevate with normal aging and in RA (37,38,39,40,41). Aging has been proposed as mediated by systemic, chronic, low grade inflammation [6] and RA is a typical of diseases mediated by a systemic, chronic inflammation. The serumpentosidine level may bea reflection of chronic inflammation of unknown etiology. And most anti-inflammatory /anti-rheumatic treatment for RA decreased serum pentosidine [41,42,43,44].

There may be possible reasons why anti-dyslipidemic activity of both fenofibrate and statins in the present long-term study was not clear in spite of the previous report thatfenofibrate improved the most lipid levels significantly in our short-term study [23]. 1) the dyslipidemia assessed by lipid profile in Japanese population was basically mild probably due to the traditional diet (45,46) and the contribution of lipid-lowering agents to the treatment may be sometimes difficult to assess. 2) To treat chronic inflammatory diseases such as RA, a combination of several drugs with different mechanisms of action that may affect lipid metabolism has been routinely used [9,10,14].

In conclusion, anti-inflammatory therapy including anti-RA drugs, fenofibrate and statins may be useful to prevent inflammation-associated cardiovascular diseases with dyslipidemia, hyperhomocysteinemia and elevated serum pentosidine.

REFERENCES

[1] Gonzalez A.,MaraditKremers H.,Crowson C.S.,Nicola P.J.,Davis J.M.III.,Therneau T.M., Roger V.L., Gabriel S.E. The widening mortality gap between rheumatoid arthritis patients and the general population.*Arthritis Rheum.* 56:3583-3587,2007.

[2] Mok C.C.,Kwok C.L.,Ho L.Y.,Chan P.T.,Yip S.F. Life expectancy,standardized mortality ratios, and causes of death in six rheumatic diseases in Hong Kong, China.*Arthritis Rheum.* 63:1182-1189,2011.

[3] Crowson C.S., Liang K.P., Therneau T.M., Kremers H.M., Gabriel S.E. Could accelerated aging explain the excess mortality in patients with seropositive rheumatoid arthritis? *Arthritis Rheum.* 62:378-382,2010.

[4] Radovits B.J., Fransen J., Shamma S.AL., Eijsbouts A.M., van Riel P.L.C.M., Laan R.F.J.M. Excess mortality emerges after 10 years in an inception cohort of early rheumatoid arthritis. *Arthritis Care Res.* 62:362-370,2010.

[5] Sokka T., Abelson B., Pincus T. Mortality in rheumatoid arthritis:2008 update. *ExpRheumatol 26* (suppl.51):S35-S61,2008.

[6] Goto M. Inflammaging (inflammation-aging): A driving force for human aging based on an evolutionarily antagonistic pleiotropy theory?.*BioScience Trends.* 2:218-230,2008.

[7] Souen S. Mortality and causes of death in RA patients. *Rheumatol.* 37:164-168,2007.

[8] Shinomiya F., Mima N., Nanba K., Tani K., Nakano S., Egawa H. Life expectancies of Japanese patients with rheumatoid arthritis:A review of deaths over a 20-year period. *Mod.Rheumatol.* 18:165-169,2008.

[9] Kaplan M.J. Management of cardiovascular disease risk in chronic inflammatory disorders. *Nature Rev.Rheumatol.* 5:208-217,2009.

[10] Nashel D.J. Is atherosclerosis a complication of long-term corticosteroid treatment? *Am. J. Med.* 80:925-929,1986.

[11] Jonsson H., Nived O., Sturfelt G. Outcome in systemic lupus erythematosus: A prospective study of patients from a defined population. *Medicine.*68:141-150,1989.
[12] Raynauld JP. Cardiovascular mortality in rheumatoid arthritis: How harmful are corticosteroids?*J.Rheumatol.* 24:415-416,1997.
[13] Urowitz M.B., Bookman A.A.M., Koehler B.E., Gordon D.A., Smythe H.A., Ogryzlo M.A. The bimodal mortality pattern of systemic lupus erythematosus. *Am. J. Med.* 60:221-226,1976.
[14] Lazzerini P.E.,Capecchi P.L.,Bisogno S.,Galeazzi M.,Marcolongo R.,Pasini F.L. Reduction in plasma homocysteine level in patients with rheumatoid arthritis given pulsed glucocorticoid treatment. *Ann Rheum.* Dis 62:694-695,2003.
[15] Symons J.D., Mullick A.E., Ensunsa J.L., Ma A.A., Rutledge J.C. Hyperhomocysteinemia evoked by folate depletion: Effects on coronary and carotid arterial function. *ArteriosclerThromb Vas.Bio.l* 22:772-780,2002.
[16] Chen YS., Yan W., Geczy C.L., Brown M.A., Thomas R. Serum levels of soluble receptor for advanced glycation end products and S100 proteins are associated with inflammatory, autoantibody, and classical risk markers of joint and vascular damage in rheumatoid arthritis. *Arthritis Res.Ther.* 11:R39,2009.
[17] Ueno H., Koyama H., Fukumoto S., Tanaka S., Shoji T., Emoto M., Tahara H., Inaba M., Kakiya R., Tabata T., Miyata T., Nishizawa Y. Advanced glycation end products, carotid atherosclerosis, and circulating endothelial progenitor cells in patients with end-stage renal disease. *Metabolism.*60:453-459,2011.
[18] Halliwell B., Hoult J.R., Blake D.R. Oxidants, inflammation, and anti-inflammatory drugs. *FASEB.J.* 2:2867-2873,1988.
[19] Taneda S., Monnier V.M. ELISA of pentosidine, an advanced glycation end product, in biological specimens. *Clin Chem.*40:1766-1773,1994.
[20] Sell D.R.,Monnier V.M. End-stage renal disease and diabetes catalyze the formation of a pentose-derived crosslink from aging human collagen. *J.Clin Invest.* 85:380-384,1990.
[21] McCarey D.W., McInnes I.B., Madhok R., Hampson R., Scherbakov O., Ford I. Trial of Atrovastatin in rheumatoid arthritis(TARA):double-blind, randomized placebo-controlled trial. *Lancet.* 363:2015-2021,2004.
[22] Staels B., Koenig W., Habib A., Merval R., Lebret M., Torra I.P. Activation of human aortic smooth-muscle cells is inhibited by PPAR alpha but not by PPAR gamma activators. *Nature.*393:790-793,1998.

[23] Goto M. A comparative study of anti-inflammatory and antidyslipidemic effects of fenofibrate and statins on rheumatoid arthritis. *Mod Rheumatol.* 20:238-243,2010.
[24] Araki A., Sako Y. Determination of free and total homocysteine in human plasma by high-performance liquid chromatography with fluorescence detection. *J.Chromatogr.* 422:43-52,1987.
[25] Sanaka T., Funaki T., Tanaka T., Hoshi S., Niwayama J., Taitoh T., Nishimura H., Higuchi C. Plasma pentosidine levels measured by a newly developed method using ELISA in patients with chronic renal failure. *Nephron.* 91:64-73,2002.
[26] Ganji V., Kafai MR. Population reference values for plasma total homocysteineconcentrations in US adults after the fortification of cereals with folic acid. *Am. J.Clin.Nutr.* 84:989-994,2006.
[27] Albert MA. Inflammatory biomarker, race/ethnicity and cardiovascular disease. Nutr Rev 65:S234-S238,2007.
[28] Lin CP., Chen YH., LeuHB., Lin SJ., Chen YL., Huang SL., Chen JW. Anti-inflammatory strategies for homocysteine-related cardiocascular disease. *Front Biosci.* 14:3836-3845,2009.
[29] Oudi ME., Aouni Z., Mozigh C., Khochkar R., Ganiueni E., Haouela H., Machghoul S. Homocysteine and marker of inflammation in acute coronary syndrome. *Exp.Clin.Cardiol.* 15:e25-e28,2010.
[30] Dierkes J., Westphal S., Luley C. Serum homocysteine increases after therapy with fenofibrate or bezafibrate. *Lancet.* 354:219-220,1999.
[31] Giral P., Bruckert E., Jacob N., Chapman MJ., Foglietti MJ., Turpin G.Homocysteine and lipid lowering agents. A comparison between atorvastatin and fenofibarate in patients with mixed hyperlipidemia. *Atherosclerosis.* 154:421-427,2001.
[32] Luc G., Jacob N., Bouly M., Fruchart JC., Staels B., Giral P. Fenofibrate increases homocystinemia through a PPARα-mediated metabolism. *J.CardiovascPharmacol.* 43:452-453,2004.
[33] Hoelstra M.,Haagsma C.J.,DoelmanC.J.A.,van de Laar M.A.F.J. Intermittent rises in plasma homocysteine in patients with rheumatoid arthritis treated with higher dose methotrexate. *Ann. Rheum. Dis.* 64:141-143,2005.
[34] Van Ede A.E., Laan R.F.J.M., Blom H.J., Boers G.H.J., Haagsma C.J., Thomas C.M.G., de Boo T.M., van de Putte L.B.A. Homocysteine and folate status in methotrexate-treated patients with rheumatoid arthritis. *Rheumatol.*41:658-665,2002.

[35] Rho Y.H., Oeser A., Chung C.P., Milne G.L., Stein C.M. Drugs used in the treatment of rheumatoid arthritis:relationship between current use and cardiovascular risk factors. *Arch. Drug Inf.* 2:34-40,2009.

[36] Verma S., Szmitko P.E., Ridker P.M. C-reactive protein comes of age. *Nature ClinPrct. Cardiovascular Med.* 2:29-36,2005.

[37] Miyata T., Ishiguro N., Yasuda Y., Ito T., Nangaku M., Iwata H., Kurokawa K. Increased pentosidine, an advanced glycation end product, in plasma and synovial fluid from patients with rheumatoid arthritis and its relation with inflammatory markers.*BiochemBiophys Res. Comm.* 244:45-49,1998.

[38] Rodriguez-Garcia J., Requena JR., Rodriguez-Segade S. Increased concentrations of serum pentosidine in rheumatoid arthritis. *Clin Chem.* 44:250-255,1998.

[39] Ozturk H.S.,Cimen M.Y.B., Cimen O.B., Kacmaz M., Durak I. Oxidant/antioxidant status of plasma samples from patients with rheumatoid arthritis. *Rheumatol. Int.* 19:35-37,1999.

[40] Sulman ME., Heimburger O., Barany P., Anderstam B., Pecoits-Filho R., Ayala ER., Oureshi AR., Fehrman-Ekholm I., Lindholm B., Steminkel P. Plasma pentosidine is associated with inflammation and malnutrition in end-stage renal disease patients starting on dialysis therapy. *J. Am. Soc.Nephrol.* 14:1614-1622,2003.

[41] Ramasamy R., VannucciSJ., Yan SSD., Herold K., Yan SF., Schmidt AM. Advanced glycation end products and RAGE: a common thread in aging, diabtes, neurodegeneration, and inflammation. *Glycobiol.* 15:16R-28R,2005.

[42] Kageyama Y., Takahashi M., Nagafusa T., Torikai E., Nagano A. Methotrexate reduces the levels of pentosidine and 8-hyroxy-deoxy guanosine in patients with rheumatoid arthritis. *Mod.Rheumatol.* 17:398-402,2007.

[43] Kageyama Y., Takahashi M., Ichikawa T., Torikai E., Nagano A. Reduction of oxidative stress marker levels by anti-TNF-α antibody, infliximab, in patients with rheumatoid arthritis. *Clin. Exp.Rheumatol.* 26:73-80,2008.

[44] Urios P, Grigorova-Borsos AM, Sternberg M. Aspirin inhibits the formation of pentosidine, a cross-linking advanced glycation end product, in collagen.*Diabetes Res.Clin.Pract.* 77:337-340,2007.

[45] Nakashima M., Sakurai M., Nakamura K., Miura K., Yoshita K., Morikawa Y., Ishizaki M., Murakami K., Kido T., Naruse Y., Sasaki S., Nakagawa H. Dietary glycemic index, glycemic load and blood lipid

levels in middle-aged Japanese men and women. *J.Ather.Thromb.* 17:1082-1095,2010.

[46] Inadera H., Hamazaki T. Cholesterol controversy: Cutoff point of low density lipoprotein cholesterol level in guideline by Japanese Atherosclerosis Society.*Jap. J. Hygiene.* 65:506-515,2010.

INDEX

B

C

E

F

I

J

K

L

M

N

O

P

Q

R

S

T